The Essential Guide to Time Travel
Temporal Anomalies & Replacement Theory

M. Joseph Young

ISBN: 978-1-989940-29-7
Copyright © 2021 Mark Joseph Young
Dimensionfold Publishing
Prince George BC CAN

Acknowledgments

I must thank:

Dimitrios "Jim" Denaxas and his Dungeons & Dragons™ friends, whose questions about time travel in Terminator and other films forced me to think about the subject

E. R. Jones, who by involving me in Multiverser™, the game in which everything is possible, caused me to put the first ideas to paper and to launch the Temporal Anomalies in Popular Time Travel Movies web site

Fans of that website, and particularly **John "A1Nut" Cross**, who pressed me to continue expanding it and exploring time travel issues

My five sons, and particularly **Kyler**, who so often served as sounding boards for these ideas

Eugene Whong, Producer and Cohost of The Time Travel Forensics Podcast, whose critical comments on the third draft were invaluable

My wife Janet, who has put up with all the writing and efforts to publish for most of half a century

And of course **God the Father and the Lord Jesus Christ**, who put me on this completely unanticipated path.

Contents

Preface	1
how this book came to be.	
List of Citations: Time Travel Movies	3
so the reader can avoid spoilers.	
List of Citations: Other References	5
The Core Theories	7
fundamental ways time travel is handled.	
Fixed Time Theory	10
that the past cannot be changed.	
Multiple Dimension Theory	11
parallel and divergent universes.	
Replacement Theory	12
the ability to alter history.	
Fixed Time Theory Examined	15
details and problems of the theory.	
The Predestination Paradox	18
loops with uncaused causes.	
Becoming Your Own Grandfather	21
a particular predestination paradox problem.	
The Grandfather Paradox	23
the reverse problem, preventing your existence.	
The Novikov Self-Consistency Principle	30
a mathematical argument for fixed time.	
Clarifying Fixed Time	31
immutable means immutable.	
Multiple Dimension Theories Examined	33
more than one history of everything.	
Types of Parallel Dimensions	33
what we might expect.	
Unparalleled	35
how time travel unravels the theory.	
Types of Divergent Dimensions	37
a different way to the same outcome.	

The Two Brothers Problem in Multiple Dimension Theory	38
a simple logic problem that complicates things.	
The Temporal Duplicate Problem with Divergent Dimensions	45
if the traveler repeats the same trip.	
Other Problems with Divergent Dimensions	49
including thermodynamics.	
Replacement Theory Examined	54
real time travel with free will.	
The N-Jump	57
the preferred outcome of time travel.	
The Infinity Loop	59
the ultimate temporal disaster.	
Sawtooth Snaps and Cycling Causalities	61
repeatedly changing timelines	
Where the People Go	63
explaining what happens to everyone when time ends.	
Niven's Law	65
uncreating time travel.	
Temporal Duplicates and Replacement Theory	69
objects and people doubled by time travel.	
Rate of Change	70
when does the change in the past alter the future.	
The Spreadsheet Illustration	73
demonstrating the anomalies mathematically.	
The Butterfly Effect	80
small changes can have big impacts.	
The Genetic Problem	82
how the entire population of the world can be changed.	
Analyzing Examples	86
applying temporal theory to time travel stories.	

Analyzing Back to the Future	89
showing a film that got most of it right.	
The Beginning	90
reconstructing the original history.	
Changing History	92
how Marty altered his own past.	
Quibbles	93
all the little problems.	
Another Change	96
the other version of Marty.	
An Alternate Explanation	101
applying alternative theories.	
Analyzing Terminator	104
reconstructing the analysis of the first film studied.	
A Fixed Time Solution	105
looking at the story if time is immutable.	
Other Dimensions	107
a consideration of whether Multiple Dimension Theory works here.	
Rewind, Replace	110
the Replacement Theory solution.	
Ratcheting	115
a sawtooth snap.	
Analyzing Los Cronocrimines, a.k.a. TimeCrimes	117
studying a cleverly complicated story.	
The Final History	119
the version of events presented to us in the film.	
An Original Timeline	123
what happens to Hector when no one comes from the future.	
The First Time Traveler	125
what is wrong with the actions of the second Hector.	
The Second Time Traveler	126
why the third Hector controls the others.	
The Girl in the Woods	130
how she got there originally.	

The Woman on the Roof	135
who fell to her death originally.	
The Third Hector	138
how the final history is created.	
Analyzing Predestination	142
unraveling a challenging paradox by popular demand.	
Temporal Order	142
putting the events in a final history.	
Birthing Baby Jane	147
how Jane becomes her own parents.	
Original History	149
who Jane was before time travel.	
Becoming Barkeep	157
the next transition.	
Meeting Yourself	165
what happens when the time traveler encounters himself.	
Jumping Into Bodies	174
a specific trope of some time travel stories.	
How to Change the Past	177
a workable method of using time travel to alter recent events.	
Foreseeing the Future	182
stories that look like information traveling from the future but which are better understood otherwise.	
Toward Two-Dimensional Time	188
discussion of an undeveloped model for time travel.	
The Perpetual Barbecue	199
a short story built on Replacement Theory.	
About The Author	209

Preface

I suppose the ironic part is that at least compared to many of my readers I'm not really that much of a time travel fan. Oh, I had read Wells' *The Time Machine* by the time I was twelve, and *Time Tunnel* was one of my favorite television shows back then. But I'm not more fond of time travel than of a lot of other science fiction and fantasy. I never watched *The Sarah Connor Chronicles*, and—well, maybe I put so much effort into understanding temporal anomalies that I was too judgmental of the efforts of popular entertainment. I stopped watching *Star Trek Voyager* before the end of the first season, after they had hit me with three completely impossible temporal disasters. So why do I put so much effort into this?

I think it started with *The Terminator*. I'm not even sure why I had seen it, but my *Dungeons & Dragons*™ players for some reason expected me to be able to explain it. So I did, and that got me thinking deeply enough about the events in that film, and in *Back to the Future*, that I began forming a theory, a way to understand what happens if a person travels to another point in time.

Then I was asked to help create *Multiverser: The Game*, and as part of that it was at least possible that time travel would be involved. I wound up including the basic framework of the theory, and as the game was published I decided that one way to help promote it would be to create a web site discussing the time travel ideas and applying them to popular movies. This became very popular, and the site kept expanding to cover more movies and answer theory questions along the way. I gradually became something of a respected authority on the subject of time travel. People would write to me with questions; college professors would include my web site in their syllabi.

More than once I was asked if I was going to compile it in a book, and I would just point to the massive somewhat disjointed *Temporal Anomalies in Popular Time Travel Movies* web site[1] and ask how. But then Ken Goudsward wrote and said he was interested in publishing several of my books, and particularly in having me compose something about time travel. We both agreed that something that presented and explained temporal theory interspersed with examples from movies might work, and so this book came to be.

Because quite a few movies are cited and there are significant spoilers involved in using them as examples, the next page will provide a list of films you should watch (if you haven't already) before you read about them in the text which follows. Hopefully I managed to list them all. I have marked with asterisks those for which there are stronger reasons for you to have viewed them before reading about them here, usually because their use in the text would constitute a significant "spoiler" or because understanding the text would be difficult without having seen the film. Following that is a list of other cited sources, but these are less significant in terms of your need to be familiar with them. Both lists are in the order in which they are first mentioned in the text of the book.

---

[1] Formerly at GeoCities, currently at http://www.mjyoung.net/time/.

List of Citations: Time Travel Movies

Back to the Future*
Terminator*
Los Cronocrimines (a.k.a. TimeCrimes)*
Predestination*
Minority Report
Next
12 Monkeys*
Somewhere In Time*
The Final Countdown
La Jetée
The Philadelphia Experiment
Paradox
Synchronicity
Back to the Future Part II*
Source Code*
Terminator 2: Judgment Day
Primer
The Butterfly Effect (and sequels)
A Sound of Thunder
O Homem Do Futuro (a.k.a. The Man from the Future)
When We First Met
See You Yesterday
Terminator 3: Rise of the Machines
Star Trek IV: The Voyage Home
Timerider: The Adventure of Lyle Swann*
The Time Machine*
Bender's Big Score
Frequency*
Déjà vu
Men in Black III
Back to the Future Part III
Millennium
Flight of the Navigator
Bill and Ted's Bogus Journey

Timeline
The History of Time Travel
Looper*
Meet the Robinsons
Kate and Leopold
H. G. Wells' The Time Machine
Terminator Salvation
11 Minutes Ago
Donnie Darko
Bill & Ted's Excellent Adventure
About Time
Edge of Tomorrow
Mr. Peabody & Sherman
Timecop
Harry Potter and the Prisoner of Azkaban
Peggy Sue Got Married
Hot Tub Time Machine
Premonition
Groundhog Day
12:01
Watchmen
X-Men: Days of Future Past
Time Lapse
Mirage

Detailed analyses of most of these movies appear on the web site:
Temporal Anomalies in Popular Time Travel Movies, http://www.mjyoung.net/time/

List of Citations: Other References

Dungeons & Dragons, role playing game
Multiverser: The Game, role playing game
Temporal Anomalies in Popular Time Travel Movies, web site
The Time Travel Forensics Podcast (formerly Café Seoul Podcast)
The Time Machine, book by H. G. Wells
Time Tunnel, television series
The Sarah Connor Chronicles, television series
Star Trek Voyager, television series
Rip Van Winkle, short story by Washington Irving
Buck Rogers, character featured in short films, television, and other media
Sliders, television series
All You Zombies, short story by Robert A. Heinlein
Ladyhawk, fantasy movie
Job, a book in the Bible
Ringworld, setting for several books by Larry Niven
The Mote in God's Eye, novel by Larry Niven and Jerry Pournelle
Red Dwarf, television series
Tikka to Ride, episode of Red Dwarf
A Critique of the Spreadsheet Theory of Time, web page
A Sound of Thunder, short story by Ray Bradbury
Star Trek, television series
Tomorrow is Yesterday, episode of Star Trek
The Magicians, television series
Dr. Who, television series
Mawdryn Undead, episode of Dr. Who
Quantum Leap, television series
The Dragon and the George, short story and novel by Gordon R. Dickson
Dark, television series
7 Days, television series

Dune, book trilogy by Frank Herbert
Time Patrol, a collection of short stories by Poul Anderson
Inferno, episode of Dr. Who
RPG Review, web site
Multiverser: The Second Book of Worlds, role playing game supplement book

The Core Theories

There are a lot of time travel stories, many of them rendered to film, and so you might expect that there would be many theories about time travel. There are of course a lot of minor variations, but in essence there are three core theories. What is perhaps the most respected is known as *Fixed Time Theory*, the notion that you cannot change the past, and if you become involved in the past you will only discover that you were always involved in the past and didn't change anything. The second we should call *Multiple Dimension Theory*, which comes in two major forms both of which attempt to resolve temporal problems by moving the traveler to a different but identical universe. The third has become known as *Replacement Theory*, which is probably the one most used in fiction, in which the time traveler makes changes to his own history.

We will begin with an overview of each of these, and then delve more deeply into the advantages and problems of each.

In considering the logic of time travel, the first issue to address is the nature of time itself. Two principle forms are logically plausible: either all of time exists in some static form from beginning to end (or eternity to eternity), or only the present moment exists, the past lost and destroyed, the future not yet created. Either view is defensible. However, from a time travel perspective, if only the present moment exists, time travel is impossible nonsense: you cannot go to a place that does not exist. Thus we can have the *Rip Van Winkle*-type story exemplified by *Buck Rogers* in which the character sleeps or is in suspended animation and emerges in the future, and stories like *Minority Report* and *Next* in which the character is able to predict future events, but not one in which a character travels to the past. Time travel of that sort seems to require that something like a "timeline" exists, that the past is still the present "somewhen", and the future is "somewhen" to which we can travel.

Some avoid the "timeline" with a rather more complicated "timeplane", that is, movement along many parallel timelines which are staggered along the same events. By moving laterally across time, you travel to what appears to be the past or future but is actually a separate universe lagging or leading our own. Thus the future and past of our universe do not exist, but are duplicated in other universes also without futures or pasts. We cannot travel to our own past, but we can travel to something indistinguishable from it, and change it with impunity because it is not our universe. This will be considered further in discussing Multiple Dimension Theory.

The very concept of a "time line" suggests that time is a dimension—not in the sense of another universe, but in the sense that length, width, and height are dimensions. Some insist that this is not true, that time is different in kind from these; yet despite the differences, time has the same impact on reality as spatial dimensions, other than that it is what we can call the dimension of change: absent time, everything would remain the same, because change implies time, a "before" and an "after".

We can see that time is like a spatial dimension by means of a simple thought experiment. The corner of the desk is a point in space; if you place a book there, it occupies that point in space, and you cannot place another book in the same place. Suppose that the book is actually an object with no dimensions, which occupies a point to the exclusion of all other objects. In space with zero dimensions, that is all that can exist. If, though, we take the edge of the desk to be the first dimension, we see that we can place other books along the edge, but only one at that point that we call the corner. However, if we then take the perpendicular edge as defining a second dimension, the surface of the desk becomes a two-dimensional space, and we can place a second book at the same point along the first edge by placing it at a different point along the second edge. It is thus in the same place in one dimension by being in a different place in the other. We can repeat this by adding a third dimension, stacking the second book on top of the first, at which point both books are in the same place, on the corner of the desk, in the two dimensions that define the desktop, by being in a different place in the third.

We could hypothesize a fourth spatial dimension, such that two books can be in the same place on the corner of the desk touching the desk by being in different places in this fourth spatial dimension;[2] but we can also arrange for the two books to be in exactly the same place in those three dimensions by removing the first and replacing it with the second. We then have an object in the same place in three dimensions because it is in a different "place" in the fourth. Time is thus a dimension similar to the spatial ones, and arguably could exist very like a line or fourth axis on a graph.

In order for time travel to be possible, time must be a dimension very like this. The questions are whether we can move to other points along it (which is assumed by the concept of time travel), and whether it is mutable or fixed.

---

[2] It was suggested that I provide an image of this to clarify it, but a two-dimensional illustration of a four-dimensional reality is probably not helpful even if someone has the talents to make it possible. I searched the Internet for four-dimension graphs, and the images were not particularly useful.

Fixed Time Theory

Some think Fixed Time Theory is the "scientific" view, and there are many scientists who believe that this is the correct theory, and many ideas about how and why it works. In essence, this says that you cannot change the past.

What adherents to this theory usually miss is that it is far more deterministic than they suggest. If you cannot change the "past", you cannot change the "future" either.

Suppose you were to travel from 2020 to 2010. According to this theory, your arrival in 2010 was always part of the history of the world—there never in any sense was a version of history in which you did not arrive in 2010. But whatever you do, time is going to unfold such that 2020 will be exactly as you knew it, because you cannot change the past.

What you're missing, though, is when you are in 2010, 2020 is the future. It is in a sense your past—your sequential past, since you have already lived in that year—but for everyone else in 2010 it is the future, and as far as they know it has not yet happened and will be formed based on their choices. Yet if you cannot change 2020 because you cannot change 2010, then their choices are fixed, destined, and they cannot change 2020 either. That, though, means that if you are alive in 2020, 2030 is also already fixed, and whatever you think you are choosing of your own free will is actually your predetermined choice, because you cannot change 2030 either. The only difference is you don't already know anything about what happens in 2030, so you can't recognize that you are not free to form it.

That's not a pleasant idea for us—but that doesn't mean it isn't true. There's a lot more to consider about this theory, but that's the core of it: all of history has been predetermined from the origin of the universe to its demise, and we cannot change it, only experience it as it unfolds.

It is not easy to find story examples of Fixed Time, not so much because people don't attempt to make them but because in almost every case the same events can be explained under a different theory—particularly when one of the popular tropes such as the *predestination paradox* is included. Some movies popularly thought to be examples of fixed time (and we will return to at least some of these to show alternative explanations) include the original series entry *The Terminator*, *12 Monkeys*, *Somewhere In Time*, *The Final Countdown*, *La Jetée*, *The Philadelphia Experiment*, *Paradox*, and others that will be mentioned in connection with later issues.

Multiple Dimension Theory

As mentioned, there are two primary varieties of Multiple Dimension Theory. The popular name for this, Parallel Dimension Theory, only describes one of them accurately; the other, probably more popular in fiction, is better dubbed Divergent Dimension Theory.

Parallel Dimension Theory maintains that our universe is part of a multiverse of sorts, in which there is a vast, some say infinite, number of universes already existing. For time travel purposes, these are all identical, and so when the time traveler leaves for the past he lands in a different universe that is indistinguishable from his own, and so whatever changes he makes affect that universe, and not the one from which he originated. There are other versions of this. In the television series *Sliders* it was assumed that every universe was different from the one from which the travelers departed, even if only in seemingly insignificant ways (such as the replacement of the Golden Gate Bridge with the Azure Gate Bridge). The movie *Synchronicity* appears to follow a similar theory. There are scientists who believe there is evidence for the existence of these alternate dimensions, but what they are like is not something we can demonstrate.

According to Divergent Dimension Theory, there is originally a single history of the world and thus a single universe up to the moment someone arrives in the past. The traveler's arrival in the past causes the creation of a new history that diverges from the original one, and thus a new universe. Again, nothing the traveler does impacts his own history, and the original history of the world remains unchanged for everyone still in it. He is in a separate universe which he created in which everyone and everything has been duplicated and is experiencing an altered history.

It is not always easy to distinguish one of these theories from the other, but movies that clearly use some form of Multiple Dimension Theory include *Back to the Future Part II*, *Source Code*, and *Synchronicity*.

Replacement Theory

The essence of Replacement Theory is that a time traveler has complete freedom of action in the past, and can change it—but there are consequences. There are quite a few variations on those consequences, which we will explore.

In discussing replacement theory there are a number of conventions used for identifying timelines. In essence, there is an original "AB" timeline, in which "B" represents the moment in time from which a traveler departs for the past, and "A" the moment in time at which he intends to arrive. Since his arrival in the past alters history by virtue of the fact that he was not there in the original version, this creates the "CD" timeline, "C" the same instant in time as "A" but different because of the presence of the time traveler, "D" the same instant as "B", the moment of the departure from the future. If the anomaly is complicated, there might be an "EF" timeline and a "GH" timeline and others beyond that, each different from all those preceding it.

The most common outcome of time travel under this theory is probably the *infinity loop*, time trapped forever between two alternately repeating histories each of which erases itself and causes the other. The most desired outcome is the *N-jump*, in which the altered timeline causes itself, and history can continue into a future. Either of these might be extended by intervening histories in what is called a *sawtooth snap* or *cycling causality*, in which several different histories are chained, each causing the next, possibly perpetually but more typically until either an infinity loop or an N-jump results.

Identifying examples of Replacement Theory movies is almost a silly task, since most time travel movies are about how someone from the future changed the past, and many Multiple Dimension Theory stories and nearly all Fixed Time Theory stories can be explained by Replacement Theory. However, a few of the obvious films in this regard are *Terminator 2: Judgment Day*, *Back to the Future*, *Primer*, *The Butterfly Effect*, *A Sound of Thunder*,

O Homem Do Futuro (a.k.a. *The Man From the Future*), *When We First Met*, and *See You Yesterday*.

There are movies that don't fit into these categories. Some of them raise temporal issues but don't actually involve time travel, or are best understood as not involving time travel. Some do involve time travel but have no coherent theory of time, mixing tropes from different theories into a hash of temporal nonsense.

Each of these theories has advantages and flaws and variations, and so we will have to look at them individually in more detail.

Fixed Time Theory Examined

Fixed time theory views all of history as already completed; time is not exactly an illusion, but the way we experience it is. It might best be understood by speaking of a road that runs from your house to the store, and of your travel along it. The road exists; the store awaits ahead. Your movement does not cause the store to come into existence, but only brings you to it. In the same way, your death (for something that is ahead for at least most of my readers) stands at a specific moment in the future, and you are approaching it. It has in some sense "already" happened; you just have not yet reached the moment where it exists. You feel as if you are making choices that bring you forward in life, but in some sense you are only following the road, acting out what you are destined to do. Thus if you are going to travel to the past, your departure from the future has already occurred and so has your arrival in the past, and everything which from your perspective you are going to do you have from another perspective already done. Thus you cannot change the past, because the past is already written; you cannot change the future, either, because that, too, is already written.

Unless a time traveler goes to other universes time must be something like a dimension, and thus something we might call a timeline exists from the distant past through the present. We might ask whether the line continues into the future, but that is already part of the more basic question: is it fixed, or mutable?

Advocates of fixed time theory maintain that time is immutable, that the past cannot be changed. It also means that the future cannot be changed, either. It is easy enough to suppose that everything James Cole does in the past in

12 Monkeys he was destined to do in the past because he had already done it, but what is overlooked is that for the young James Cole watching in the airport, everything he is going to do in the future is equally destined. Thus when the fixed time theorist says that you cannot kill your own grandfather because we already know that you did not do so, implicit in that is that you have no control over your destiny, that all your future choices are already made, that you are fated to a life you have not yet chosen but cannot avoid choosing.

Most of us intuitively balk at this idea that we have no free will; it seems as if we do. That, though, is an inadequate basis for concluding that we do. It is certainly arguable that we will do what we, by heredity and environment, are programmed to do, and that this feels like freedom because all the constraints are internal, within our own personalities. It is similarly possible that the entire universe is so programmed, that just as billiard balls deflect at predictable angles everything in the universe will follow a predetermined sequence of events. The problem is that once time travel is introduced you have the possibility of an instability in the programming process. What is to prevent the program from looping with the command $A=A+1$, or $-A=A$, for causes in the future to have effects in the past that alter those causes in the future? Fixed time advocates assert that this cannot happen, but the basis for that assertion is circular, that if it did, it would change the past, and since you cannot change the past, whenever you travel to the past you begin a trip you have already completed.

Yet even given a deterministic model of human choice, it is difficult to imagine that no trip to the past would ever alter events in a way that would change the factors determining those choices. If on Monday you ate a hamburger and got sick, on Tuesday you might send a message to yourself that says "don't eat the hamburger", and if having received the message you chose not to eat the hamburger and you did not get sick, then on Tuesday you would not send that message. Fixed time advocates assert that you could not do this—either you could not choose to send the message, or having received it you could not choose not to eat the hamburger. On the other hand, if having turned right on Broad Street you later discovered that there was a terrible accident to the left on Broad Street which you avoided, you could send yourself a message that said, "don't turn left on Broad Street," and your previous self could receive that message and obey it, because that does not change what he would have done anyway. At this point, it appears that the universe, or whatever it is that prevents certain willed acts but not others, has sufficient intelligence to foresee what acts in the future will alter the past and which will not. It suddenly seems to become a matter of divine intervention, of God under a pseudonym preventing catastrophe.

To the fixed time advocate, though, time is not at all like a sequence of choices, but like a puzzle that fits together only one way, and we discover how it fits as we live through it. The events in our lives are like tiles in a mosaic being constructed by someone else, but that there is no one doing the construction and no one planning the pattern, and yet somehow it all fits properly.

Fixed time thus poses some intellectual problems in regard to choices and future events which it tends to gloss. It creates a world in which I cannot help writing this book and you cannot help reading it, because we are destined to do what we do not know we have already done in the future. We have no control; it is an illusion. We cannot even control whether or not we believe in fixed time.

The Predestination Paradox

Fixed Time storytellers like to create causal loops, in which events in the future are caused by events in the past which were in turn caused by those events in the future. *12 Monkeys* wants us to believe that the virus was released in the past because time travelers from the future put access to it in the hands of someone who was expecting an apocalypse. Perhaps the most brilliantly crafted of such stories is *Predestination*, from a Robert Heinlein story in which the central character is both of his own parents, who are the same person, and he is his own mentor and the villain he seeks to stop.

The concept is usually illustrated and defended with the example of a billiard ball which rolls into a wormhole, comes out at a different point sometime in the past, and collides with itself. In doing so it should knock itself off its original trajectory such that it will not enter the wormhole. But obviously if it does not enter the wormhole, it will not exit the wormhole, and so will not interfere with its own path. The problem of the causal loop is reduced to its simplest terms.

Scientists working with this model have demonstrated that it is always possible to create an outcome in which the billiard ball coming out of the pocket in the past will in colliding with itself knock itself into the entry. They have further asserted that since such a solution always exists, it must be that this will always happen, for "nature" will not permit an uncaused result. Thus having demonstrated that it could happen, they assert that it must.

There are several logical fallacies here, but the most glaring one is the confusion of the possible with the necessary. One could as easily extrapolate that because the billiard ball entered the wormhole in the original timeline, when it exits the wormhole it will never collide with itself. This is as valid an assumption as the other, and any ninny can devise a thousand models in which the collision doesn't happen for every one in which it does. More to the point—and a problem with the fixed time theory generally—is that it begs the question of how this occurred in the first place. Basic Newtonian Mechanics will tell us that if the collision is necessary to drive the billiard ball into the wormhole, then if the collision does not occur there is no way that the ball will ever enter the wormhole. That is, if we have one entrance to the wormhole and one exit, and we adduce a model in which the ball coming out of the exit will collide with itself such that it will drive itself into the entrance, then we know intrinsically that if the collision does not happen, the billiard ball will never enter the wormhole initially, and so won't exit in the past. Occam's razor should alert us immediately that the solution to this problem is that the billiard ball never enters the wormhole at all, because there is nothing to drive it that way. No, the fixed timeline theory fails because it is easier to suppose that time travel will not occur than to suppose that having occurred in the future it will cause itself in the past. If it must have happened in order to happen, it is easier to suppose that it never happened. Arguably, anything that only happens if it happens, which ultimately causes itself, never happens.

Paradox aside, let me make this clear: if the billiard ball is rolling toward the wormhole in the first place, but it collides with itself coming out of the wormhole, it will be deflected such that it will never enter the wormhole; but if it is not bound for the wormhole initially, there is no reason for us to assume that it will come out of the wormhole and deflect itself into it. The only thing that can be said is that a billiard ball coming out of a wormhole in its own past will not collide with itself under any scenario, and that patently requires an intervention of a divine level.

To quote a movie that has nothing to do with this, "I believe in miracles—it's part of my job."[3] However, I do try to base my science on natural laws, not divine interventions.

Does this mean that such loops are inherently impossible? It seems that they are impossible under Fixed Time, and adherents to that theory disagree with each other about this, some arguing that as long as all causes and effects are included in history the loop can exist, others that such a loop could never come into existence. Under Replacement Theory, though, it is possible for an initial cause to begin the events and the time traveler to interfere so as to become the replacement cause. For example, in order for Skynet to target Sarah Connor in *The Terminator*, some version of Skynet must have come into existence, and we know that that only happens because Cyberdyne obtains parts from the terminator sent back to kill her. We get the solution to this in *Terminator 3: Rise of the Machines*, as a different but similar Skynet is launched by the United States Air Force, providing us with the artificially intelligent enemy that ultimately sends the first terminator back to start the loop. Similarly, in *Somewhere In Time* Elise McKenna fell in love with a man who left promising to return, and decades later she mistook Richard Collier for him, and told him to come back to her. He then manages to travel back to her time and place, and interferes with her meeting with her original love. In these ways such self-supporting loops can be launched and maintained.

---

[3] The author is Chaplain of the Christian Gamers Guild, and author of *Why I Believe*, Dimensionfold, 2020. The quote is from *Ladyhawk*.

Loops that involve information are easiest to support. In *Star Trek IV: The Voyage Home*, Mr. Scott trades the metallurgical formula for transparent aluminum for a sheet of plexiglass. In a passing explanation, Scotty says, "How do we know he didn't invent the thing?" However, this means that in this timeline no one discovered transparent aluminum; the knowledge was passed from future to past, and it simply was. Dr. Nichols is credited for it, but was given the formula. Under fixed time, this is a logical possibility, that it was neither invented nor discovered but the knowledge passed from future to past. The process of discovery is not necessary.

Becoming Your Own Grandfather

One classic form of the Predestination Paradox has the time traveler becoming his own grandfather. This happens literally in *Timerider: The Adventure of Lyle Swann*, in which he does not know he has gone back in time, meets the woman whom he does not know is his grandmother, and leaves her in the past when he is rescued. Fixed time adherents treat this as an ordinary predestination paradox, but it is much more complicated than that.

Using *Timerider* as an example, Lyle Swan meets a woman in the past whom he impregnates, and she gives birth to his son who is his father. Lyle is his own grandfather. If we resolved this using replacement theory, there must have been an original Stranger who met and impregnated Grandmother, fathering Father. Father's DNA is 50% Stranger and 50% Grandmother. He marries Mother, siring Lyle, 50% Mother, 50% Father—which makes him 25% Stranger, 25% Grandmother. Lyle then disrupts the meeting between Grandmother and Stranger, fathering Father. Father's DNA is now 62.5% Grandmother, 25% Mother, and 12.5% Stranger. Lyle's becomes 62.5% Mother, 31.25% Grandmother, and only 6.25% Stranger. But Stranger is the only contributor of a Y chromosome here, so as his percentage falls to 3.125% of Father and 1.5625% of Lyle, and then <1% of Father and <0.5% of Lyle, eventually one of them will become a woman, and the chain will break.

Fixed time ignores this genetic problem, supposing that if you are your own grandfather history has always been so.

There is a similar but different problem with physical objects in such loops, illustrated by the medallion passed from Lyle Swann to his grandmother and later inherited by him, and by the pocket watch Elise McKenna gives to Richard Collier and later receives from him in *Somewhere In Time*. Using the watch as an example, we first see elderly Elise McKenna give it to young Richard Collier, who takes it to the past and leaves it with young McKenna, who later gives it to Collier when she is old. We have an object with no origin in time. Fixed time theory claims this is acceptable, and replacement theory puts the origin of the watch in the lost original history. But unlike the information passed from Scott to Nichols (in *Star Trek IV: The Voyage Home*), the watch ages. When McKenna hands it to Collier, she has owned it for sixty years; he then owns it for a decade before giving it back to her. When she got it, then, it was at least seventy years old—but when he got it, it must have been 130, and thus 140 when he gave it to her, increasing seventy years with each pass.

Fixed time expositions ignore this aging factor; the watch does not age, because if it is there it must be there.[4]

The Grandfather Paradox

Although the predestination paradox includes the notion of a time traveler becoming his own grandfather, there is a separate paradox, the reverse of that, known as the grandfather paradox, in which the time traveler kills his own grandfather before the grandfather has children, thus undoing his father's birth and from that his own. Not having been born, the time traveler cannot travel to the past and cannot kill anyone, and so he cannot prevent his own birth. It of course need not be a grandfather specifically, nor need it be intentional; what matters is that the traveler has undone a link in his own ancestry.

It is interesting that the problem is always phrased with a grandfather. Rarely does anyone speak of traveling to the past to kill himself, even though that is the practical outcome of the scenario. This phrasing, though, permits fixed time theorists many options in explaining how it is that your efforts did not result in your own undoing. They would assert first that we know you failed, because had you succeeded you would never have been born to make the attempt. If pressed—how could you fail if you

---

[4] It might be suggested that when traveling to the past the watch "youthens", and so does not decay to dust. However, if it happens to the watch it should also happen to Collier, and were he to become seventy years younger he would be completely non-existent.
Alternative explanations include that the watch is somehow replaced, presumably outside the knowledge of either of its owners and without being visibly different than they remembered, or that by some means akin to divine intervention the watch itself is indestructible. Again, we are stepping outside of science for solutions.

detonated a nuclear device under his bed while he was in it?—it shifts to whether you killed the right person. In one way or another, they insist that you cannot kill your true grandfather because history is fixed, and your birth demonstrates that your true grandfather lived to sire your father.

This answer is entirely unsatisfactory for some theorists, who cannot understand how the concerted efforts of untold numbers of time travelers to undo their own lives the easy way could be thwarted by nature, as if it were omniscient (because it can always predict paradox) and omnipotent (because it can always prevent it). The fixed time theorist, though, relies on the determinism inherent in that theory, that everything that ever has or ever will happen is already established, and thus whether we travel to the past or stay where we are we cannot really change anything.

To explore this further, I am going to suggest that perhaps I might wish to commit suicide. Suicide is not easy, and it is threatening. There are few means, if any, which are certain and painless, and for many the consequences of failure are great.

I could eliminate much of the trouble with suicide by a simple method I've imagined. I could travel back to the beginning of the last century and locate my grandfather, and kill him before he ever has the chance to sire my father. This would not be a terribly difficult task. I know where he lived; he was my grandmother's next door neighbor. He was also seventeen years her senior, so I have plenty of time. They were in a small town in rural Mississippi, so there's not too much danger of losing him in the crowds. If I've got access to a time machine, a few weapons of mass destruction should not be problematic. I can take Mississippi off the map, if need be. That should finish him.

Note how perfect this plan is. If I succeed, not only do I end my suffering, I will have eliminated all that suffering I've already experienced. Suicidal people at least since Job tend to see the misery in their lives overbalancing any good that ever happened; the good is merely a deception, a moment when the reality was disguised. Thus for the suicidal person, there is nothing worth saving of the past; to wipe out the entire existence is the best course. I can do it. I can make it so I was never born.

Further, the consequences of failure are minimal. Probably no one would ever know I made the attempt. I can't suffer brain damage, disfigurement, crippling. Even social disdain is unlikely.

I don't have to worry about who might find the body; there will be no body in this century. I need not fear the pain, as there will be no pain. In all, it is the perfect suicide plan.

To plans like this, fixed time theorists will only say, you will fail. Something will go wrong. After all, they argue, clearly you are alive, so clearly you have already failed to kill your grandfather. Perhaps your weapon will misfire. Perhaps you'll accidentally kill the wrong person. Maybe you'll be apprehended before reaching your target. It is possible that you will relent upon seeing him, and be unwilling to cause that pain to someone else. Maybe you'll just miss. Somehow, you will happen to fail.

This sounds so reasonable. After all, they aren't saying that I can't do it, only that I won't. I have failed, so I will fail. We don't know what will go wrong, but something will go wrong.

Obviously, I am not the first person to have thought of the idea of suicide by time travel. I will not be the last. Some will read the idea in this book; far more have read it elsewhere, or encountered it through other media. There must be thousands of people already who wished they could kill a parent or grandparent before the chain of conception which led to them was initiated. If we assume that the universe is unending and populated by uncounted intelligent beings,[5] there must be billions of creatures who have an idea not much different from this.

It is also generally true of technology that once it is discovered it grows more common. At one time automobiles were toys of the rich; now in many parts of the world they are basic necessities. Television broadcasting began in corporate hands; today your local church or civic group could, for not too much money, own its own low power television (LPTV) station, if it could program it, and many cable companies have public access channels. Not so long ago state medical boards were trying to control rising health care costs by limiting the number of CAT scan machines; today such machines are considered essential to the operation of all hospitals and emergency medical facilities. Who even owned their own computer fifty years ago? Today we all carry them in our pockets. Thus it is reasonable to suppose that once a working time machine is built, it will be just a matter of time before time travel becomes widely available.

---

[5] I don't actually believe that the universe is infinite either temporally or spatially, or that it is populated with uncounted intelligent creatures. Those might be so, but I consider the evidence stronger against them. This is a hypothetical example.

We must now contend with the twin facts that billions of people want to undo their own births by killing their grandparents and that time travel is within their grasp.

This suggests that there will be at least millions of people traveling back in time, taking with them advanced weaponry that makes modern thermonuclear devices look like firecrackers, trying to erase their own existences by destroying their ancestors, with no regard for the collateral damage whatsoever.

The fixed time theory tells us that every one of these highly motivated, capable, prepared, and equipped individuals will just happen to fail.

People who defend this notion get offended when I tell them that what they're saying is God will not allow it to happen. I don't mean it to offend; it is what they are saying. Of course, they're trying to be scientific, so they don't say God. Usually they say Nature would prevent it. Yet it is a strange thing, this Nature that does not allow paradox. Somehow Nature must be able to perceive when Paradox is going to occur. After all, no one has said that time travelers wouldn't be able to move about in the past as freely as they do in the present; it is only claimed that any action which would create paradox would be prevented. Thus Nature must somehow be able to spot actions which are likely to create paradox, so she can prevent them. I get letters all the time from people who can't spot actions that are going to create paradox. I even see such actions in movies that were very expensive to make and paid someone a lot of money to try to produce a credible story that presumably wouldn't contain paradox. It's not always so easy to see when an action might create a paradox, and I'm surprised this Nature, whoever she is, is able to do so before the fact.

This Nature also must have incredible power, to be able to prevent all those events. Millions of capable individuals are going to have their specific intended actions thwarted by her efforts, and it will be done so subtly there will be no hint it was anything other than happenstance. A gun misfired. A bomb did not go off. You killed the wrong person. The police caught you first. Nature is a very clever lady, if she can do all that.

The fact is, if you're talking about an entity which is intelligent enough to spot potential paradox before it occurs, and capable enough to prevent it, you're describing some notion of a deity. It might not be the God I, as a theologian, would describe, but it is a very powerful and wise deity who obviously has some benevolence toward creation to take such action. I don't care what you call it, you mean God. You mean it won't happen because God would not allow it.

I am, as I say, a theologian by training. I'm no expert in the field, but I have taught undergraduate studies. I believe in God, and in divine intervention, and in providence. However, I'm far too knowledgeable to think it safe to assume that we can do any fool irresponsible thing we want, and God will protect us from the consequences. It doesn't work that way in our lives as individuals, and it's never been that way in the world at large. People thought God would not allow a disease to be so deadly and devastating as the Black Plague was. Some thought God would not allow man to fly, and others that He would not allow us to reach the Moon. There were those who thought the atrocities of Hitler were beyond what God would permit, and those who honestly believed that God would not let us split the atom. All of these people were wrong.

Now we have a theory that is no more than *God will not allow us to alter history*. I don't see how this is any more likely to be the limit than any other thing someone believed He would not permit.

I suspect there may well be divinely-appointed limits to what we can do; I also recognize that we have no way of knowing what they are. It is irresponsible to suggest that we do know something so occult.

In the end, there are only two possibilities. Either God will prevent us from changing the past, or one of those suicide bombers will get through. If it's the former outcome, we will probably find time travel itself to be impossible. If it's the latter, the Fixed Time Theory will crumble in a heap as the world scrambles for some other theory of time which resolves the disaster that has been created—but that since history was changed and no one knows the original history, we probably will never know that it was changed.

What is the point?

The Fixed Time Theory does not resolve paradox. It ignores it. It makes the unfounded claim that paradox cannot happen.

There is no evidence that the past is immutable. That is at best a hypothesis and at worst wishful thinking at this point. We cannot assume that it is impossible to damage something we do not fully comprehend. If the past proves mutable, we must be prepared for the consequences of changing it.

The Novikov Self-Consistency Principle

Igor Dmitriyevich Novikov has contributed to the theory of time travel with what is called the *self-consistency principle*, or sometimes the *self-consistency conjecture*. In its simplest form, it says in essence that time travel is only possible if it does not cause a paradox. In that sense, it is a restatement of fixed time theory, in which somehow the universe knows what actions in the past will cause a paradox and chooses to prevent such actions. What makes Novikov significant, though, is that he is a physicist who has worked through a mathematical solution to problems in time travel under relativity, in which time travel is possible under these conditions.

That does not mean that the principle has been proved. It only means that there is a possible mathematical model for the universe in which time travel is possible given this principle. There are other known solutions to the equations which do not allow time travel, and it is assumed that additional solutions exist. However, given the billiard ball example, in which a billiard ball coming out of a wormhole from the future knocks itself into the wormhole, it is asserted that there are always many possible trajectories for this, and therefore it must be that it follows one of those trajectories.

What is unclear is what mechanism prevents an outcome that does not fit the theory: given that I cannot travel to the past to kill my own grandfather, why not? Apart from that it would cause a paradox (as the chief morlock incongruously asserts in the remake of *The Time Machine*), what prevents a time traveler from changing the past? It is simple enough to devise a time travel task that would be unresolvable and yet not difficult to perform. Novikov asserts that because a paradox is mathematically anomalous it cannot happen. Yet the point of a temporal paradox up front is that it appears to be impossible—the predestination paradox his theory permits not less so than the grandfather paradox it excludes. That what appears logically impossible on its face should also prove mathematically impossible under a certain model of the universe is not at all surprising. It may do more to call into question the model itself than to tell us anything about the universe.

Thus in the end, Novikov has said that if time is unalterable, it cannot be altered. The question is whether the premise is correct.

Clarifying Fixed Time

Writers often get confused concerning what it means for time to be immutable. For example, in *The Time Machine* (the remade version) Emma dies from a gunshot from a mugger, and then Alexander Hartdeggen travels back four years to save her from the mugger, only to have her die hit by a car whose brakes fail. Hartdeggen commiserates that he cannot change the past—but he did change the past. He did not save Emma, but instead of her dying from criminal assault she died from product failure.

Were we to apply fixed time theory to this circumstance, Hartdeggen could not have prevented Emma from being shot. Under fixed time, small changes are as significant as large ones. We have to accept that nothing changes.

Does it really make a difference?

In the first scenario just presented, there is somewhere a mugger who has now killed someone. Presumably police are seeking him, and have an eye witness who can probably identify him. If he does not flee the city they will probably find him, even in late nineteenth century New York, and he will probably be convicted of murder and likely executed. If he does flee the city, there will be a small decrease in the number of muggings in New York and probably a small increase in muggings in another city, perhaps Newark, or Philadelphia, or Hartford, or Boston, as the mugger takes his business elsewhere.

In the second scenario, it is probable that someone else was accosted by the mugger that night, which is going to impact the life of the new victim. The mugger will continue to practice his craft in New York. Meanwhile, the owner of the vehicle is going to be traumatized by the accidental death of a woman he never met, hit by his own car when he left it unattended outside a store and his brakes failed. We can only guess at the psychological results of this.

Further, if the mugger killed Emma, the mugger fled with the engagement ring; if she died from the car hitting her, Alexander has it. So where is the ring four years later when Alexander is building the time machine?

When we discuss Replacement Theory we will address the *butterfly effect*, which is critical to a theory in which time is mutable. For the present, though, it should be understood that immutable time means exactly that, that not even the smallest detail can be changed.

Multiple Dimension Theories Examined

For those who object to the notion that everything is determined and you cannot change history, it has been proposed that you can change history—you just cannot change *your own* history. To achieve this, the theory proposes that any travel to the past takes you to a different universe. Of course, it's a bit silly to talk about more than one "everything", but this holds that there are alternate dimensions, worlds like our own which exist in a timespace we cannot perceive.[6]

There are two distinct conceptions of this, each with its own variants and problems. One is that these alternate dimensions have existed since the beginning of time; for clarity, this is called *parallel dimension theory*. The other maintains that originally there was only one history of everything, but that certain events, notably the arrival of a time traveler from the future, cause the universe to split, creating a distinct separate history; again for clarity, this is called *divergent dimension theory*.

Types of Parallel Dimensions

Some scientists have performed experiments with lasers and photons which they maintain demonstrate the existence of other dimensions. Whether these other dimensions are actually distinct universes in any way similar to ours or are instead spatial dimensions of our own universe which we are unable to perceive (for example, something akin to hyperspace) cannot be

---

[6] Somewhere there is a list of killer exam questions, one of which reads, "Define the universe. Give three examples."

determined from the evidence, but it does make the existence of such universes a possibility.

Assuming such universes exist, we have three hypotheses concerning them.

The first is that they are all different. This is exemplified in the television series *Sliders*, in which the titular characters are moving not through time but from universe to universe, each universe different from home sometimes by a little, sometimes by a lot. However, it is unclear whether the show represents the universes as different enough. There might be universes with a different number of spatial dimensions, universes in which there is no earth, or the earth is still a ball of magma or has no atmosphere or no breathable atmosphere. If such alternate universes exist, this is perhaps the most plausible form for them. It is not really relevant to time travel other than that if this is reality travel to other universes would not appear to be time travel.

The second is that these parallel universes are all exactly identical until altered by an outside force. This is difficult to distinguish from divergent dimension theory, but is distinct at the fundamental level that the second universe has always existed independently of the first. Most time travel stories that assume original parallel universes assume this model, and assert that when you travel to the "past" you land in a different universe, and so any changes you make are made to that other universe, not to your own, avoiding the paradoxes connected to time travel.

The third configuration of parallel universes eliminates the need for "timelines" by placing each universe between two others a fraction of a second apart. Time travel in this configuration moves across universes. That is, if you attempt to travel back an hour, you will move across dimensions to that universe whose present time is exactly one hour behind yours. Again since you are in a different universe anything you change will not impact your own history, because you are in a different world. If you return to the future, you will again cross dimensions in the other direction, to the one that is an hour ahead of the one from which you departed, a different universe from the one you changed.

Unparalleled

Implicit in the theory of time travel to parallel dimensions, that is, those which have always existed, is the belief that such parallel universes are exactly like our own in every detail until a time traveler arrives and alters them. Thus each individual dimension has its own fixed time, with all the inherent determinism involved, but without the problems of free will causing paradox because we are not interfering with our own past. A grandfather paradox is not a problem because if you kill the person who would be your grandfather in this universe, it has no impact on your real grandfather who is alive in your universe. A predestination paradox is also not a problem, because if you marry the woman who would have been your grandmother, her descendants do not then include you, because the woman who actually is your grandmother is in another universe married to your grandfather, who is not you.

More complicated, though, is that it is assumed that when we travel to another universe it is identical to our own until we change it. This fails a simple logical consideration. To begin, there is either an infinite number of such universes or a finite number; either way, the odds are against us being the first to discover time travel, and the greater the number of universes the worse those odds become. Some hold that such universes are temporally linked, such that if you travel a hundred minutes into the past you cross to that universe a hundred minutes behind your own, with ninety-nine universes intervening at one-minute increments. (Minutes are used here for convenience; whether they are at steps or on a continuum is irrelevant, but steps are easier to grasp.) Let us then suppose that Traveler goes from our universe, "H" for home, one minute to the past to universe "I", and he interferes with events in universe "I" such that his alternate self in "I" does not in turn travel to "J". That means that the history of "I" is now different from the history of "H", but since no traveler arrived in "J", its history will be the same as "H" and its Traveler will travel to "K", changing the history of "K". This single time travel event thus alters the history in half of all existing universes, between those from which a time traveler departed and those to which a time traveler arrived. Yet if we are not the first universe in line, this has already happened to our universe and to all the universes adjacent to us. Therefore the parallel universes which would theoretically be identical until altered by a time traveler will already have been altered by many time travelers before we arrive (as probably ours will have as well, although we might be ignorant of that fact), and we will not have the predicted experience of being in a universe entirely like our own.

Of course, not every version of parallel dimension theory has them so linked temporally; but that part is irrelevant except to make the image simpler. If a dimension traveler arrives in some other dimension and prevents his own departure from it, it divides all histories equally into two versions; and if it does not so divide all histories, then we assume that a time traveler from some other dimension could arrive in ours and prevent our traveler from departing, but at the same time not do so because our traveler departs—exactly the kind of paradox this theory was supposed to resolve.

The only way to avoid this is to demand that the same things happen the same way in every universe; but in that case you have not solved the problems of fixed time, because even with an infinite number of such universes extending infinitely forward and back through time, once they are all linked such that the same history occurs in each, you merely have fixed time on a larger scale.

Types of Divergent Dimensions

As there are three types of parallel dimensions, there are in a sense two types of divergent dimensions.

One could be described as *all possible universes exist*. That is, whenever there are two possibilities, "A" and "B", there is a universe in which "A" happens and a different universe in which "B" happens. Of course, many events are not binary, that is, there would also be universes for possibilities "C", "D", "E", and on through the alphabet. The number of universes thus increases rapidly, and also exponentially, as for one event we might have "A", "B" and "C", while independently of that we have another event which has possible outcomes "x", "y", and "z", and so we have universes in which the events are "Ax", "Ay", "Az", "Bx", "By", "Bz", "Cx", "Cy", and "Cz", nine distinct universes from just three possible outcomes of two different events. Given the sheer number of events happening at any instant and the number of possible outcomes, the number of different universes would be beyond calculation.

Some attempt to limit this by asserting that only the choices of volitional creatures create new universes. That is, the question of whether or not there will be a landslide at a particular time on a particular day is effectively fixed by physical laws such that this will happen in every universe in which nothing else differs to change that. The question of whether a particular person is caught by that landslide depends on choices that person makes, and thus will create different universes.

The other type of divergent dimensions eliminates the volitional element as well, that is, much as with Fixed Time Theory what people (or animals) choose to do is predetermined such that they could not do otherwise. In this view, there will be only one universe, only one history of the world, until someone invents time travel and alters the past. At the moment that the time traveler arrives a new universe is created. The old original universe still exists, in which the time traveler leaves from a point in the future but never arrived in the past, and a new universe diverges from this as a time traveler arrives from the future. There is thus only one universe up to the moment of the arrival of a time traveler, at which point the presence of the time traveler causes the creation of a new and different universe.

The Two Brothers Problem in Multiple Dimension Theory

The Parallel and Divergent Dimensions Theories are repeatedly presented as "this is how time travel really works", and these are very interesting theories—*but they're not time travel.*

This illustration may explain why.

You had a brother. He was two years younger than you, and you were always very close growing up. However, when you were a high school sophomore, your brother contracted a rare, debilitating, and fatal disease, and it was clear that your time together in this life was limited.

You gave up your dreams of becoming an architect, and instead picked up more biology, went on to college and into medical research. In ten years, you managed to devise a cure for that disease, a medicine which would have saved your brother's life.

It's too late, though. Your brother died mere months prior to the development. Besides, it would have required more years of testing before the FDA would have approved human trials, and the drug is only truly effective if administered in the early stages. Your brother has died; there's nothing you can do about it.

The stubbornness that caused you to press on and develop this cure so swiftly won't let you accept this. You return to graduate school, pick up the needed background, and earn a doctorate in physics. Then you assemble a research team, get grant funding, and devise a method of traveling to the past. You build your machine, a mere ten years after you began your researches in the field, a mere twenty years after your brother was diagnosed with this dread disease. Still wanting to recover your lost brother, you take the cure, samples, formula, and research information, and travel back thirty years. You know who the leaders are in several pharmaceutical companies, and you arrange for one of them to take on the testing and approval of the drug. All is going well; you return to the future, the time you left.

But to which future have you returned? Multiple dimensions theory proposes that there are now two futures—the one you know, and the one you have created. What is the history of the world when you return to your own time?

Most theorists preferring this theory maintain that you can only return to the future you created; therefore, we'll examine the minority view first, that you return to the world you left.

It is a given of the Multiple Dimensions Theories that you don't change your own past. Thus if you return to the world from which you departed, you will find it unchanged.

That means all your labor has been vain. Your brother died a decade ago, and neither your medicine nor your time machine was able to save him. Everything that drove you to do this has failed. Your brother is dead.

Proponents of this theory will say that certainly this is so; it's not a problem. However, consider this: there is no evidence that time travel works, but your own claim to have been in the past—a claim that is unverifiable, because nothing you did in the past was ever done. Your project is shut down; you lose your funding, no one publishes your papers, and no one is interested in giving you another job. Time travel doesn't work, the science headlines say. You cannot travel to the past. As Al in *Quantum Leap* addressing the Senate Committee, you have not even a scrap of evidence that your time travels are anything other than your fanciful stories told to get more funding, and you don't even have the faint hope that your changes to history will work their way through to the present and save your project. You were never in the past—the real past, the only past that any normal person means when they say "the past". You were in another universe which was remarkably similar to this one, but that's not the same thing.

As mentioned, the majority of those who support the Divergent Dimensions Theory believe that if you return to the future after traveling to the past, you will always arrive in the universe you created. Let us look at the history of this other world.

You delivered your wonder drug to the pharmaceutical company, and since they didn't have to spend a fortune in development they went forward with testing. The drug proved itself in early tests, and was approved for clinical trials, and by the end of the decade was in common use. Thus when your brother got sick, the doctor said how very fortunate he was. Ten years ago what he had would have been fatal, but today there's a wonderful treatment for it, and he should recover fully. He does so.

Of course, he's not exactly *your* brother. There's a fifteen-year-old version of you in high school when he's diagnosed, hoping one day to be an architect. His brother contracts some disease which was once serious, but now is easily cured. He's happy for his brother, and for the wonders of modern science; but he's got none of the motivation which so controlled your life. He goes on to study architecture, and becomes a great architect. He travels all over the world, designing buildings, monuments, cities, airports, bridges, and so many other structures; and he comes home regularly to see his brother, whom he is only dimly aware how fortunate he is to have. By the time he's thirty-five, and his brother thirty-three, he is known around the world for his architectural skills.

By the time you were thirty-five, your brother had been dead a decade, you had developed a cure for a killer disease and a cogent theory of time travel, and left for the past. You know nothing about architecture, but what you knew as an interested high school student.

You travel back to the future, to the year when you were thirty-five, and go to find your brother, the one whose life you saved. Yet he is not your brother; or perhaps more precisely, you are not his brother. His brother is an architect, world famous in that field. His brother knows nothing of medicine, pharmacology, physics, or time travel. You're an impostor; you're not even all that good an impostor. You know nothing that either your brother or his brother have done over the past twenty years. You are unfamiliar with the fields of study his brother pursued. Given the very different lives you've led—yours constantly cooped up in labs, his often on construction sites—you probably don't even look much alike. Sure, you've got the same DNA and fingerprints, but do you really want to push the issue that far? (Not even clones have the same fingerprints, although they share DNA.) You are not this person's brother. You're a stranger, and you will always be a stranger.

Meanwhile, back in your own world, your brother died. You did not save your brother; you saved someone else's brother.

What's worse, now you've died. Oh, you don't realize it—you're still alive as far as you know. But you stepped into that quantum leap accelerator and vanished, and there was never another trace of you anywhere in time or space. You never appeared in the past; nothing you were going to do to leave your mark in the past was done; you never returned. You will be reported killed in a failed time travel experiment. The experiment will be labeled a failure, and shut down. It will be remembered as a famed folly, with your name attached. The conclusion for all the world will be that time travel doesn't work.

All of this is rather fanciful, and shows up many of the problems with Multiple Dimension Theory. In the end, no matter what happens, those of us at the starting point will believe time travel is impossible, and will give it up. It doesn't work, because traveling to another universe is not time travel.

But the story just told will never happen, even if Multiple Dimension Theory is true. The world doesn't work that way.

It makes for wonderful fiction to imagine the lone inventor working in his lab to create the next great discovery; and I would not minimize the work of lone inventors. They are, however, mainly a nineteenth century phenomenon. Dr. Frankenstein's search for the secret of life, Dr. Jeckyl's effort to unlock his inner self, Alexander Hartdeggen's invention of the time machine—these are all the works of wealthy nineteenth century self-financed aristocratic inventors, of which the nineteenth century offered many. Today's inventors don't discover science; they apply it as technology. Science is advanced in large, well-funded laboratories by teams of researchers working under the guidance of a few who have a proven track record of successful research and careful experimentation. No one is going to build a time machine and assume that it works without testing it; no one is going to suddenly leap back in time, any more than we sent people into space before we were certain we could send empty space capsules up there, and domestic test animals.

The first time traveler would not be human; it would not be animal, nor even vegetable. It would be a carefully crafted chunk of material, a very pure element, most likely carbon but a number of elemental metals also commend themselves. It would be crafted precisely, probably a cube or possibly a sphere of specific dimensions and mass. It would be sent back a very short temporal distance, probably mere minutes. But it would never arrive in the past, and so all experimentation would grind to a halt over this one problem.

You see, everything that is said of the consequences of sending a human back in time must equally be true of the consequences of sending a cubic centimeter of carbon back in time. Time travel can't work one way for people and a different way for things. And until it has been demonstrated that a block of pure carbon can be reliably moved from the present to the past, no one is going to allow a human being, nor even an earthworm, to make a trip in any such machine.

The ultimate problem with multiple dimension theory is that if it works it can't be demonstrated. Thus time travel dead-ends under this theory, as no one can know whether those who vanish are in another physical world, or have merely had their atoms scattered across the cosmos.[7]

---

[7] It has been suggested that a linked chain of identical universes resolves this, if universe A sends to universe B, and B to C, and so on. It only complicates the issues, as universe A will be a sender but not a recipient, and so will abandon time travel experiments, and on the next experiment universe B will not be a recipient, and so with each additional experiment another universe will have its experiment "fail". The proposed solution to this is that universe Z, or whatever is the last

Multiple Dimension theory is interesting, but it's not time travel. It isn't time travel unless the traveler arrives in his own past. That means he can impact his own life; and that means we need a theory that addresses what happens when someone impacts his own life.

Both the Multiple Dimension and the Fixed Time theory start from the assumption that the past cannot be changed, and so to look for a way to make time travel possible without changing the past. Multiple Dimension theory does this by saying you aren't in your own past; that in essence says that when you travel to the past, you don't travel to the past but actually do something else.

The Temporal Duplicate Problem with Divergent Dimensions

Under most theories of time travel there is the possibility of the *temporal duplicate* or *doppelganger*, a younger or older version of the time traveler. If you were to travel to the past, there would be the possibility that you might meet yourself, your former or younger self, in the past; it might also happen that were you to travel to the future you might meet your future or older self. Further, you might find that the self you meet is not you—that is, that this meeting never happened for the older version of you when he was the younger version. Such other selves are called temporal duplicates or doppelgangers, and are another form of temporal anomaly which occurs in time travel.

---

universe, sends to the first. The problem at this point is you no longer have many universes, but one, and what the time traveler does to change the past is changing the past in his own universe through the chain. It won't be noticed at first, but it will be devastating eventually.

Under fixed time theory, the resolution of such an anomaly is relatively simple: if the meeting occurs, it will have occurred for all versions of the traveler. If you did not meet your older self when you were younger, you will not meet your younger self when you are older. It is not something to be avoided; it is something that cannot be caused to happen.

Parallel and divergent dimension theories also treat the matter simply: you exist in each such universe, and if your other self has not also left for yet another universe you might meet him and change his life significantly, but since he is not really you, this does not matter except in that you are demonstrably an alien visitor to this universe. However, temporal duplicates create a peculiar complication for divergent dimension theory.

It is often suggested that the simple solution to time travel problems is that the traveler creates, brings into existence, an alternate dimension, leaving the original universe unchanged but having the new one diverge from its history at the point of arrival. This appears to have been the logic of *Back to the Future Part II*, where Doc explains an image which appears to show a new history diverging from an old one;[8] some suggest that this is the solution to the problems in *Primer*. There is, however, a problem with the theory which is seen in the shifting expectations of those who hold it: what happens if the same traveler makes the same trip to the same point in the past?

Let us assume that Joe leaves from 2010 and arrives in 2000, where to avoid confusion he gives his name as Moe. He interferes with the 2000 Presidential election in Florida, resulting in an Al Gore victory (the reverse of

---

[8] This is not consistent with the other films, but the series is not entirely consistent with itself.

what happened in *Bender's Big Score*). Moe undoubtedly knows that he has effected this change; but Joe, his counterpart in this dimension, has no way of knowing either that Bush would have won or that he is responsible for the change.

It is generally agreed that Joe does not have to travel from 2010 to 2000 under this theory, because Moe is already there. The chief complication at this point is that there are two of him, Moe ten years older than Joe. The changes Moe made are made, and Joe's failure to act will not change that.

The complication is, what happens if Joe also leaves from 2010 to 2000?

The first and obvious part of the answer is that he also creates a new dimension, in which he arrives. However, from there it becomes complicated. What is the present history of this new universe?

At this point, we must consider why Joe made the trip. Here are some possibilities.

1. Joe had reason to believe that Bush would have won had someone not tampered with the Florida elections, so he traveled back to prevent that. He expects to find Moe (whom he does not know is himself). Certainly if he arrives a moment after Moe does, he would arrive in his own history and Moe would be there. Yet if Moe's acts in history are not dependent on Joe's arrival, then even if Joe arrives before Moe, Moe should arrive, because Moe's arrival is part of Joe's history, and thus a fact in the past of the dimension from which this

dimension diverges. When Joe encounters Moe, he will realize that it is he.

2. Joe had reason to believe that he was Moe, and thinks that either replacement theory or fixed time theory holds that he must make this trip and do what he already did. He does not expect to find Moe, because he suspects that he is Moe, and indeed upon his arrival he finds that he is Moe, and so does what Moe did.

The complication is that aficionados of divergent dimension theory will use whichever of those results fits the story. That is, if Joe expects to face Moe, he discovers that Moe is a parallel version of himself, and if Joe does not expect Moe to be there Moe is not.

The logical objection is that it cannot be both ways, or rather, it cannot be whichever way we wish; it must consistently be one way or the other. The logical solution is that once Moe has reached the past, his counterpart Joe cannot travel to any universe in which Moe did not or does not arrive on schedule. Whatever process somehow duplicates the entire universe and all events within it must also duplicate that arrival. Thus if Moe's arrival prevents Joe from departing, there will be two of him of different ages in that universe; but if it does not prevent Joe from departing, the next universe will have three of him, two of whom have traveled from the future, and with each additional departure there is an additional arrival in the past, each of whom comes from a slightly different version of history.

Thus although divergent dimension theory resolves many issues in stories in which the traveler does not again return to the past, it only complicates stories in which the same traveler makes the same trip, or in which the traveler attempts to "return" to his life in the new universe, occupied by his divergent self.

Other Problems with Divergent Dimensions

The salvation of Divergent Dimension Theory is also its problem: each trip creates a new universe diverging from the previous one. Although this faces most of the objections to Parallel Dimension Theory, it avoids one: travelers never go to the same universe, but to one that did not exist prior to their arrival.

Universe creation is a problem for physicists: this must create a universe as large as our own, but conservation of matter and energy declares this impossible, as it would require the energy in all matter and energy in this universe.[9] However, in most stories the mechanics of time travel are secondary to those of the effects, and this theory manages effects well. The biggest problem with this theory may be in the laws of thermodynamics, specifically the conservation of matter and energy, that matter and energy are neither created nor destroyed. You can turn matter into energy and energy into matter; but where do you get all the matter and energy to create a new universe? Some assert that this is a function of the time travel device, but in practical terms it would require the

---

[9] Some have suggested that a divergent dimension could be what some have called a "pocket universe", a dimension not larger than that which can be perceived by the time traveler. However, we must assume that the new universe contains all the people known to the traveler, who are real people, and thus it contains all of their knowledge, and all the people known to them along with their knowledge, and ultimately the created divergent universe must be as large as is known to everyone in the world, and thus as large as the original universe, or else it has to be an illusion, not a real universe at all, with no real people.

conversion of all the matter and energy of the present universe in a perfect conversion to create the matter and energy of the new one.

This is sometimes resolved through a version of *Schrödinger's Cat*, saying that matter and energy are not really configured in just one way but in many ways of which we perceive only one. Other dimensions don't really have their own matter, they have ours, arranged differently. This takes us into quantum theory, and the fact that some physicist at some point proposed that the reason we did not know whether any particular unstable atom had decayed is that at any given moment it existed in both the decayed and the undecayed state in different universes, and when we checked we would determine in which universe we were by establishing whether it was decayed or undecayed; until that moment, it existed in both states in our universe. Another scientist named Schrödinger thought this absurd, and so created a "thought experiment" to demonstrate the absurdity: place a single unstable atom in a box with a Geiger counter which on detection of atomic decay will release cyanide gas into the box, in which there is a cat who will die instantly. Since the molecule has both decayed and not decayed until we check it, the cat in the box is both alive and dead until we check. Obviously, Schrödinger argued, the cat is either alive or dead; it is not both. Therefore the multiverse theory about atomic decay is incorrect.

However, theorists rallied to the side of the original proposal: indeed, the cat is both alive and dead. Then, from this very limited and controlled starting point, there developed a notion that all creatures are both alive and dead, that everything that might happen is happening in some universe, and so the number of universes is increasing exponentially. Every imaginable universe and most unimaginable ones are real.

It is noteworthy that Schrödinger created the illustration of the cat to show how absurd such an idea is, but those who embrace the idea embrace the illustration as not at all absurd.

As mentioned, some forms of this theory suggest that the universe is constantly dividing at every point where anyone makes a choice or anything could happen or not, dividing into one universe for each event that could have happened. If you could go somewhere or stay home, there must be one universe in which you go and another in which you stay. The number of universes so created is staggering. There are at this moment hundreds of things you could do—go to bed, eat, read this page, read another page, rob a bank, assault the person in the next room. You are doing one. Multiplying the number of universes by the things you could do, even that meager list makes six universes. A second person with six choices makes the number of universes thirty-six; a third makes it two hundred sixteen—increasing exponentially with each additional person. The population of Earth is billions, with innumerable choices for each, recurring each second (or less) (because you could change what you are doing now, or now, or now). Thus the number of universes expands at an inconceivable rate.

More challenging, though, is there are at least as many wicked terrible harmful selfish things anyone could do as good wonderful helpful selfless things, but the theory multiplies by one universe for each thing. Nearly everyone alive could have assaulted someone several times today, but most did not. This theory means there must be many universes in which most people did assault someone today, and tomorrow ours might be such a universe. All possible universes existing, ours is extremely unlikely, and almost certain to turn toward the barbaric tomorrow; yet it never does. This seems fatal to such a multiverse theory, which makes for fun fiction but does not fit reality.

Sideways time sometimes applies to Divergent Dimension Theory much as it does to Parallel Dimension Theory, but it is more common in divergent dimension stories to assert that such travel is impossible. Divergent Dimension Theory then resembles Replacement Theory, in that the universe that was is inaccessible to those who left it. It is sometimes unclear in stories whether the original history has been replaced or the travelers have been isolated from it.

Travel to the future under this theory is more confused. In most versions, such travel takes the traveler to the future of the world he created; however, in some versions any travel through time creates a new dimension, and in others a traveler will return to his own universe if he travels forward the same temporal distance that he traveled back. The mechanism for this is unclear, failing to answer what happens if the traveler heads for a time in the future of his place in the past but the past of his originating point in the future. The theory does not handle forward time travel well, even when coupled with travel to the past.

In the end, while these theories of dimension travel are interesting, they are nothing at all like time travel. We need a theory that both permits a traveler to arrive in his own past and leaves him free to do what he would do even if it is not what happened in the past he knows—a theory something like that represented in *Back to the Future*, *Frequency*, *Déjà Vu*, *Men in Black III*, and other films in which time travelers change their own histories; but we need it to be logical and self-consistent.

Replacement Theory Examined

The two theories already examined share an assumption, that you cannot change your own past. Fixed Time Theory does this by fiat, that all actions in time are already set in stone and we merely experience them as they unfold, having no free choice. Multiple Dimension Theory accomplishes it by sending travelers to a different universe that is not their own history. What, though, if the assumption is false? What if you do travel to your own past, and what if that past can be changed? Most theorists see only the two theories, and think it is a choice between being unable to control your own actions on the one hand and being unable to reach your own past on the other. It is not a choice between those two options. There is a third possibility for time travel which permits people to change their own past, and addresses the consequences thereof.

Most stories use some variant of Replacement Theory. According to this theory, if you travel to the past you arrive in your own past, and what you do within that past may have consequences for your own future and the future of the universe. Thus you can change history and impact your own existence, but the consequence is that you can change history and impact your own existence.

Under replacement theory, it is assumed that no one can arrive in the past before he departs from the future, in a sequential sense, and that this sequential sequence is tied in the original history to temporal sequence. This means that there is always an original history, a version of events in which no one and nothing arrives from the future because the future has not yet been reached.[10] The time

---

[10] Astute readers have observed that at any given moment in history there is the real possibility that a time traveler has already traveled

traveler departing from the future then arrives in the past and alters that original history, in essence erasing it and replacing it with a new version. In so doing, he also undoes his own existence in the original history, and so at the moment that his departure time is reached he undoes that departure; this in turn undoes his arrival in the past—unless the version of himself in this altered history makes that same trip to the past, and so becomes the cause of his own arrival.

The theory is frequently highly rigorous, following strict rules related to maintaining causality. That is, it is certainly possible to travel to the past and kill Adolph Hitler, but you must be careful to ensure that you do not also prevent yourself from making that trip, because as easily as you can undo Hitler's legacy of horror you can also undo your own existence. Under rigorous versions of replacement theory, once you have undone your own departure to the past you have also undone whatever changes you made in the past, including the fact that you changed your own history. Time thus becomes caught in an infinity loop, something akin to a Mobius strip in time, alternating between two versions of history. Other possible anomalies under replacement theory include the *sawtooth snap*, or *cycling causality*, and the *N-jump*, each of which will be examined in detail below, illuminating the theory concepts further.

---

from the as yet further future to a point in the as yet further past, with the result that we are already living in an altered history, the original history of the world having been replaced. This is true, of course, but for the purpose of our examples we have to assume that by "original history" we mean original in relation to the current time traveler. That all of history from 1492 to 2255 has been changed by a time traveler is not really relevant to the effects of time travel jumps within that altered history.

Some less rigorous versions of the theory tend to gloss over the problem of the impact a trip has on the traveler himself. That is, once the traveler has made changes he might undo his own birth, but he does not undo his own arrival in the past. These logical inconsistencies are answered by falling back on the notion that once the past has been formed it can only be changed by someone coming from the future, and so what the traveler did in the past cannot be undone except by someone traveling from the future. In short, the only event you cannot prevent is the arrival in the past of a time traveler, even if you successfully prevent his departure from the future. Such versions of the theory begin more to resemble divergent dimension theory, in that the traveler comes from another branch of time which still exists without him; he has changed history, but has not actually changed *his* history.

Other versions of the theory attempt to limit the impact a traveler has to affecting himself, such that he has created his own history of the universe which applies only to him. Most of these are not significantly distinct from Divergent Dimension Theory, but in others the traveler builds up for himself something (sometimes called "paradox" but distinct from the use of that word otherwise) which eventually can undo his own existence as he becomes progressively more separated from reality and connected to his own in some sense unreal universe. (It is unreal because it does not exist for anyone else, and thus the people in it are not the real versions of themselves.)

The long list of movies in which replacement theory of some form seems to be in view includes *Back to the Future* and *Back to the Future Part III*, the several *Terminator* movies, *Millennium*, *Flight of the Navigator*, *Frequency*, and probably *Star Trek IV: The Voyage Home*. Many stories which appear to work under fixed time theory and some which appear to work under parallel or divergent dimension theory work as well under replacement theory.

As mentioned, under Replacement Theory there are three possible outcomes to time travel, which are themselves consequences.

The N-Jump

Of those temporal anomalies recognized under Replacement Theory, the N-jump is the good outcome, the temporal event that allows history to continue.

What identifies an N-jump is that all causes of all effects are found in a single timeline. In this, the final history in an N-jump is indistinguishable from fixed time; the difference is how it is achieved.

Under Replacement Theory, as previously stated, no one can arrive in the past before he departs from the future, in a sequential sense, and that this sequential sequence is tied in the original history to temporal sequence. Thus there is always an original history in which no time traveler ever arrived from the future.

To understand this, assume that in 2000 our traveler is ten years old, and that later in 2020 when he is thirty years old he travels back to 2000. In the sequence of events in his own life, the thirty-year-old version of him that arrives in 2000 must have been twenty-nine in 2019 before he left for the past. Yet if he has not yet left for the past, he cannot be living in a world in which he already arrived in the past. Thus there is an original history in which the traveler grew from ten to thirty years old, and then at thirty years old traveled back to when he was ten, changing the history of the world at the very least by virtue of his own arrival within it. There is then a new history of the world in which the young traveler ages from ten to thirty, unless his older self does something which interferes with that.

Since the time traveler has erased the original history and created a new version, it is logical that he has also erased his own history and thus his own existence. Since, however, his presence in the past is dependent on his departure from the future, he will vanish from the past—unless something is done which will preserve his existence in the past. An N-jump means that whatever the time traveler does in the past leads ultimately to his unaltered identical younger self making the same trip to the past, and so doing the same things. This confirms the altered history as the stable history of the world, since at the end point of the anomaly the "same person" will make the "same trip" to the "same time and place" and do the "same things".

If the traveler's duplicate self fails to confirm his actions in the past, the result will be that history has to repeat itself until it is stable. If the failure is major, such as that the traveler does not make the trip at all, this will create an *infinity loop*; a lesser failure may result in a *sawtooth snap* of some form, delaying the resolution for some uncertain number of repetitions of history. These anomalies are explained below.

Elaborate schemes have been devised by which a time traveler could change the past but ensure that he does not undo the information which led to his actions. Some of these involve temporal protections which isolate some person or group from the impact of such changes, but the better methods rely on delivering accurate information concerning the original history to the people responsible to oversee it. This will be illustrated in a later chapter on how to change the past.

Ultimately, unless the author intends to create a temporal disaster, the N-jump is the desired outcome of all time travel stories. It is the only outcome of time travel which allows a future under strict replacement theory.

The Infinity Loop

In discussing temporal anomalies caused by time travel, the *infinity loop* garners the most attention of all those occurring under Replacement Theory. This describes any scenario in which actions in one history cause a second history, but actions in the second history cause the first. It has been called a *bow tie loop*, or a *Mobius time loop*; sometimes the term *cycling causality* is used for anomalies similar to infinity loops, but the use of that term is not consistent, and will be discussed in the next section.

A simple example of an infinity loop is a case in which a time traveler attempts to prove that he is a time traveler by telling someone, "I know what you are going to have for lunch," and then reporting what he remembered from his own history, only to have the person "prove him wrong" by having something else. Thus in one history, the skeptic has a hamburger, and in the second history the traveler tells him he is going to have a hamburger, so the skeptic decides instead to have pizza in the second history; but this changes the information available to the traveler who now tells the skeptic in the third history that he will have pizza, and the skeptic instead has the hamburger he originally preferred. Since having the hamburger in one timeline will cause him to have the pizza in the next, and having the pizza in that timeline will cause him to have the hamburger in the next, history is forever trapped between these two timelines.

According to strict replacement theory, when this occurs time itself becomes trapped, and cannot progress to the next day, because the causal chain requires that the skeptic have eaten either the hamburger or the pizza, and it is impossible for either to be the final answer. Thus this is the end of time, because there can be no history beyond that point.

An infinity loop occurs if a time traveler somehow prevents his own trip to the past, such as by "fixing" what he wanted to fix. Supposing that the time traveler wished to kill Stalin before the man came to power, and in traveling to the past managed to do so, his counterpart born in the world in which Stalin never came to power will have no reason to make that trip. Yet if he never leaves from the future, he never arrives in the past, and so never kills Stalin. Thus Stalin comes to power, restoring the reason for the trip.

Fixed Time Theory resolves this by saying that it is impossible: you cannot kill Stalin because it is established that he survived, and thus that you already failed. Parallel and Divergent Dimension Theories resolve the problem by saying that you did kill Stalin, not in your own world but in another universe indistinguishable from your own world. In essence, the past you changed was someone else's past, someone who never knew that Stalin mattered, and since you came from a different universe, the fact that your counterpart never makes that trip does not alter the fact that you arrived. Thus the infinity loop is connected strictly to Replacement Theory.

This aspect of time reverting to its original form because of the failure of the traveler to depart from the future raises the strongest objections to Replacement Theory. Efforts to resolve it in a manner in which the traveler's failure to depart does not undo his arrival have led to such theories as two-dimensional time[11] and supertime.[12] Yet under strict replacement theory, the outcome is unavoidable based on causal chains, that if the cause is removed, the effect is removed in sequence, regardless of the temporal relationship.

Sawtooth Snaps and Cycling Causalities

The term *sawtooth snap* has been coined in connection with Replacement Theory. It describes any temporal anomaly in which more than two histories cause each other in sequence, such that each causes the next.

For a simple case of a sawtooth snap, imagine that on Friday morning an experimenter checks his strongbox and finds it empty, then that night uses a time machine to send one hundred dollars back in time to appear in that box that morning. In the revised history, on Friday morning he checks the box, finds and removes the money, spends one dollar on coffee, and that night sends the rest back to Friday morning. On each successive replay of that Friday

---

[11] Two-dimensional time is an undeveloped theory I put forward in response to some writings by Sergiy Koshkin, included at the end of this book.
[12] Supertime was a term presented to me by Sergiy Koshkin in a critique of my theories (published on the Temporal Anomalies web site) and later developed in an article on Gaming Outpost, since lost. It involves something like a second temporal dimension and the suggestion that a specific moment in time as defined by calendars and clocks could be in constant flux such that multiple histories exist.

the experimenter removes the money from the box, spends a dollar, and sends the remainder back to that morning. Thus each morning after the first the experimenter will find less money in the box, because each evening he sends less back. The history of the world changes incrementally with each iteration, by a decrease of one dollar in the box.

Sawtooth snaps technically have three distinct forms, distinguished by their terminations. In the example above, eventually there will be a single dollar in the box, which the experimenter will spend on coffee, and then he will send no money to the past; his counterpart that morning will find an empty box and will assume that the experiment has not yet begun, so that night he will send one hundred dollars to the past, restarting the sequence. This is called an *infinity loop termination*, because it is like an infinity loop but that it involves multiple timelines in its progression. It is possible for a sawtooth snap to resolve to an *N-jump*, a self-supporting single history in which the last timeline causes itself and time continues based on that final history; our analysis of *The Terminator*, below, contains such an anomaly. In our money-in-the-box scenario, if having spent the last dollar the experimenter sends back a note that reads, "The last dollar has been spent", and if having received that note he sends back an identical note at the end of that day, time will be confirmed into a history in which there is a note in the box in the morning identical to a note sent in the evening. It is also theoretically possible for a sawtooth snap to continue in a non-repeating causal chain, such that the end of each timeline initiates a unique new timeline (akin to the decimal string on an irrational number such as π).

The term *Cycling Causality* has also been used to describe any anomaly that resembles a sawtooth snap, but particularly those which have infinity loop terminations, in which a string of several ("2 to n") distinct histories cause each other in a repeating sequence. It is thus commonly connected to sawtooth snaps, but is a distinct term in that it includes simple infinity loops but excludes perpetual sawtooth snaps (which never repeat a duplicate history) and those which resolve to an N-jump termination. The use of this term is not consistent, though, sometimes being used to describe any anomaly that is not a simple N-jump.

Many movies are best explained using sawtooth snaps, including the first two *Terminator* films.

Where the People Go

The question is frequently raised: when a history ends, whether an original history of an N-jump or any history of an infinity loop or sawtooth snap, what happens to the people living at that moment? Do they die? Is it painful? Do they plunge headlong into some dark abyss of nothingness? The answer is difficult for some to grasp: they cease ever to have come into being. Permit this illustration.

At 8 A.M. Abe is having his usual workday breakfast of coffee and a toaster pastry. Bob is just getting out of bed, and Cal is finishing work on his new time machine. At nine, Abe punches in, Bob leaves for work, and an exhausted Cal goes to bed. Abe and Bob both have lunch around noon, while Cal sleeps. Abe goes home at 5, Bob half an hour later, and they each get dinner. Cal awakens at 7 P.M., and at eight he tests his time machine by traveling back to nine in the morning.

Abe is already at work. We can call him Abe 2, but he ate the same breakfast as Abe 1, will have the same noon lunch, leave work at five, get dinner, and by seven will be in every possible way identical to Abe 1. Technically, Abe 1 never existed, but he hasn't ceased to exist because he exists as Abe 2.

As Bob is leaving the house, his phone rings, and Cal persuades him to call out of work and meet him for brunch. There, while Cal 2 is sleeping, Cal 1 tells Bob 2 about the time machine. Bob 1 never comes into existence; that morning his life was diverted such that he became Bob 2 instead.

Then, at 8:00, unaware that Cal 1 has already done this, Cal 2 tests the time machine, traveling to nine in the morning, calling Bob, and in all ways replacing Cal 1. He alone remembers the day occurring twice, but it was the identical day (and in this case he slept through the first iteration). Bob never knew the day he went to work, and no one saw him there. Abe lived the day only once. Time continues with that as the only version of the day that was; the other has been erased and replaced.

What happens to everyone when the end of a history is reached is they return to who they were at the beginning of that history and live through the next iteration, all memory, all results of the time that passed undone. People who died have not yet died, those born or conceived are not yet alive. They did not cease to exist when their timeline ended; they returned to its beginning to live through the new version as the first time.

Niven's Law

Science fiction author Larry Niven (best known for his creation of *Ringworld* and *The Mote in God's Eye*) has published and promulgated numerous "laws" describing the universe. Among time travel fans, however, one emerges specifically as *Niven's Law*, a statement concerning time travel, which reads:

If the universe of discourse permits the possibility of time travel and of changing the past, then no time machine will be invented in that universe.

In simple terms, as popularly understood, Niven suggests that if it is possible to change the past by sending information to the past, eventually the past will be altered to an ideal state and no one will have any reason to change it further, and thus the ability to send information to the past will never be developed, there being no need for it. For example, if it were determined that the world of the future would be better if the Ford Motor Company had put an electric car into mass production in 2008, those in the future would send that information to 2008, and Ford would put the car into production; this would then change the future such that there would be no need for anyone to send that message to 2008, and the message which was received would never be sent, the world being the better version created by the delivery of that information.[13]

Two challenges may be raised against this understanding of Niven's Law, one entirely theoretical the other entirely practical.

---

[13] Of course, the traditional understanding of a time machine would involve sending a *person* to the past, but what matters for Niven's Law is that information be sent, which that person would bring.

The theoretical problem is related to causal chains. Niven's Law would seem to work in a Divergent Dimension Theory universe, in that each sending of information to the past would create a universe in which that information was received; but in a Fixed Time Theory world it does not apply at all (you cannot change the past), and in a Replacement Theory world there is a serious question as to whether an effect in the past can be maintained if it undoes its own cause in the future—the problem of killing your own grandfather, in that the cause in the future of the change in the past has been undone. In this example, since Ford put the electric car into production in 2008, no one in the future will have any reason to tell them to do so, so the message will not be sent; not having been sent, it will not be received, Ford will not put the car into production, and the world will revert to the original history in which eventually someone sends that message.

The theoretical problem in relation to Divergent Dimension Theory is that the changes to the past will always have been made in a different universe. The inventors of the time machine will never see it function, as the changes they make are actually made in someone else's past. In such universes, it is extremely difficult to demonstrate that any time machine works.

The practical problem lies in the notion that humanity in the future will be in complete agreement concerning what the best past would have been. There will always be some who believe that the world would have been better had Kennedy (or perhaps Lincoln) not been assassinated, and so might attempt to alter that event; however, as the *Red Dwarf* episode *Tikka to Ride* humorously demonstrates, not everyone will agree that such a world would have been better, and someone might well attempt to restore the original history or something like it. *Bill and Ted's Bogus Journey* similarly shows that while Rufus might think the legacy of the band has left a future paradise, De Nomolos hopes to replace it with something more to his liking.

Thus even if it is proposed that undoing a future cause does not undo a past effect (that you can kill your own grandfather) it does not follow that the world would ever reach a state which everyone agreed was the best possible history. Niven's Law, so understood, is thus probably incorrect.

As popularly understood, the law also ignores the desire to create time travel solely for the sake of knowledge. In *Timeline*, the primary use of the time machine was to investigate the events surrounding a specific time and place in history, and so to update historic records. With a machine that permits us to explore the past, historians would eagerly attempt to study innumerable aspects of history, answering the many uncertainties that plague our knowledge of ancient and even modern history. In *Déjà vu* the investigation team believes they have a device that permits them to observe events four days in the past and so investigate a crime that had already happened by watching events prior to the crime. A time machine is not worthless just because we don't want to change the past.

There is also a doppelganger problem. If Traveler goes back from 2020 to 2000 to change something and succeeds and returns to 2020, then his duplicate has no reason to make that trip and will not leave 2020 for 2000. This theory here says that this is not a problem because once the Traveler arrives in the past, he is part of the past, and he does not need to leave the future "again" to be in the past, as he is already there. Yet what if Traveler does leave from 2020 to travel to 2000, perhaps to change history from what it is (which he created) to what he does not know it originally was? Are there two versions of Traveler in the past, or only one? If there is only one, which future is he attempting to create? Thus this approach suffers from a problem similar to that of Divergent Dimension Theory, only more so: if the duplicate traveler does not depart from the future, he is still present in the past, but if he does depart from the future it makes no sense either that he does or that he does not meet himself in the past.

The only logical way for the causal chain to play requires that if the traveler made a trip to the past, the traveler's duplicate must make the same trip to the past for the same reason with the same knowledge and abilities. If he does not do so, he erases his presence in the past. This then undoes the changes he made and restores the original history, which is how we get an infinity loop.

This interpretation of Niven's Law is not necessarily what Niven meant. There are at least two other possible interpretations of it.

He may have meant that if time travel is invented and the past can be changed, it will create a temporal war in which each side keeps traveling back to prevent the other side from having time travel. Thus if the Russians invent time travel and use it to alter history, the Americans will steal the technology and travel back to prevent the Russians from discovering time travel. Then the Russians will steal the technology back from the Americans, and travel back to prevent the Americans from having it, until ultimately one side or the other manages to prevent anyone from inventing time travel, and no one has it.[14]

---

[14] This is rather cleverly the plot of the mockumentary *The History of Time Travel*.

He may have meant merely that free will is incompatible with the Novikov self-consistency principle, and therefore time travel will be impossible in any universe in which people have free will to change history. It is again wishful thinking; there is no evidence that dangerous things are impossible.

Temporal Duplicates and Replacement Theory

It is under Replacement Theory that temporal duplicates become an issue. Under this theory, there must be an original history of the world in which the time traveler did not meet himself; then the time traveler who moves to the past can meet his younger self, even though he did not when younger meet his older self. This will change his history. His younger self must then (to prevent an infinity loop) make the same trip when he is older to meet in turn his younger self, but because having met himself has changed him and he may remember the meeting, this event might be different from what he remembers. Again the younger self must make the same trip in turn. As long as each time traveler eventually makes the same trip as the original, there is a high probability that the sawtooth snap will resolve to an N-jump, and time will continue. In the final version of time, the time traveler will have played both roles in the meeting, but the "two" meetings will have been identical.

The problem is less complicated with temporally duplicated objects, but that again the "younger" of the two objects must in its turn make the same trip made by the "older" object. This means that temporally duplicated objects (or people) can exist only from the moment of the arrival of the older to the moment of the departure of the younger.

The problem is similar with travel to the future, but with an extra wrinkle. If a traveler goes to the future, in the original history he will find that he vanished from the universe at his point of departure and was never seen again; if he then returns to the past, he restores himself to the timeline, such that it is possible for his younger self coming forward to encounter him as his older self. This again alters the younger self, who must now return and become the older self for the younger self to meet, again with the hope that time may eventually stabilize such that "both" meetings are identical for the one traveler who experiences the meeting twice. Ultimately it is travel to the past that creates the duplicate, as the traveler backtracks on his own history; it is experienced differently if it begins with a leap to the future.

The issue of a time traveler meeting himself is discussed further in a later section.

Rate of Change

In *Looper*, some actions done to the younger time traveler affected the older one instantly; similarly, in *Back to the Future* (Part I) there is evidence that Marty's future is changing and affecting him directly. On the other hand, in *A Sound of Thunder* the changes come in waves like ripples on a pond, and in *Meet the Robinsons* young Lewis is racing to beat the changes happening around him (although incongruously Doris vanishes instantly). The question is, given that changing the past changes the future, how quickly do those changes occur?

One possible answer is that the changes occur instantly, that if Marty McFly prevents his own existence he immediately ceases ever to have existed. The most obvious problem with this is a sort of flicker history, the equivalent of an infinity loop oscillating between his existence and his non-existence, since of course if he causes his own non-existence he does not exist to cause it (barring Niven's Law) and so restores his existence only to undo it. How that would be experienced becomes the difficult question. The second serious problem is, as *Back to the Future* illustrates, changing your history does not necessarily mean undoing your existence; events must play through to determine how the change impacts you.

A second possible answer is that such changes move at the speed of time. Sergiy Koshkin has suggested that the changes will never reach the time traveler himself, because they propagate at the rate of time, creating an alternate history and erasing the previous one, but allowing the future to continue based on the now-non-existent version of events.[15] Thus even if the time traveler having undone his own existence will not be there to undo it so history will revert to its original form, both the original and the altered versions of time exist at different points along the timeline. The changes never reach the time traveler who made them, because he is still racing ahead of them, keeping the same lead on the changes as he created by his trip to the past. This, though, creates inconsistencies in the causal chain at least from the perspective of time travel: a time traveler who prevents World War II who then fails to prevent World War II will find that when he travels back

---

[15] Some of Koshkin's writing about Supertime has been preserved at the Temporal Anomalies web site as A Critique of the Spreadsheet Theory of Time. This early effort badly misunderstands Replacement Theory, but was written before the theory had that name.

through time he passes through some periods in which the war happened and others in which it did not.

A third solution attempts to compromise, such that the changes propagate forward through time at faster than the speed of time, but the time traveler is able to outrun them, to reach his own future before the changes do, and have the world change around him when they reach him. This has story appeal: it allows our time traveler the opportunity to fix what he did in the past, but limits the time in which he can do this. However, it is difficult to find a logical basis on which to base a rate of change that is neither at the speed of time nor instantaneous, so this devolves to fanciful conjecture.

The best solution recalls the issue of the nature of time, treating time as having a static existence experienced as if in motion. This results in a solution in which change occurs instantly but is experienced at the rate of time. If we consider all of history as a complex set of related equations, $A+B=C$, $C+D=E$, $E+F=G$, the value of G is dependent on the value of A and changes the instant A changes; but the steps of the equations occur in sequence. In this illustration, we experience the steps at the rate of time, even though the changes stretched out in the future occur instantly—using a time machine to move to the future, we would discover the altered history already awaiting us. Thus in one sense if Marty undoes his own existence, he immediately ceases ever to have existed, while in another sense having undone his own existence he continues within history until the change plays through to the new altered history. As in the film, he has time to correct the problem, partly because it is uncertain whether he has ended his own existence, but also because his birth, still in the future, and his departure for the past, still in the future, are events that have in a sense "not yet not happened". This will be discussed further in the next section.

This thus gives us a logical basis for discussing how time travel impacts history, in that the changes are immediate but not experienced until we reach the points in time when they occur. It also becomes a logical basis for handling changes to history.

The Spreadsheet Illustration

To this point time has been discussed as if it were something that moves, carrying us along with it. Some of the more astute will suggest that this is not the way time really is, if it is indeed a dimension. After all, space does not move; we move through space. So, too, time must be stationary. It is we who move.

Seeing time as stationary, and events as being in their place along that medium, actually works better. It solves a lot of the confusion about why certain events don't happen sooner. For example, why doesn't Marty McFly instantly vanish the moment he has undone his own existence by preventing his father from meeting his mother? If his entire universe was at that instant undone (as indeed we ultimately conclude it was), how can he still exist now, and yet not exist later?

One way to perceive this is by considering a computer spreadsheet program. This example will show how events can change instantly and yet not immediately; and it may help to display the nature of our three basic anomalies for those who are having difficulty grasping them in the context of rewind and replace. If the reader is somehow not familiar with the way spreadsheets work, I trust this will still be clear enough.[16]

Imagine opening a spreadsheet program—Excel is the common one for most computer users, although Lotus 1-2-3 is the one on which I first learned. You see a gridwork of what are called cells, in numbered rows and lettered columns. Thus there is a block near the top left corner of the work area which is in the first row and the first column. The first column is designated A, and the first row is designated 1, so that block is known as A1. The block below it would be A2, and the block beside it would be B1.

Imagine that we've typed a value into A1. There is now a number there. What that number is does not matter at this point.

In A2, we're going to enter a formula. A formula is a mathematical equation which calculates the value for that cell. The beauty of a spreadsheet program is that you can use in a formula the value of any other cell by designating the address of the cell, and that's what we're going to do. We're going to create a formula which makes the value of A2 dependent on the value of A1. For example, we could say that A2=A1+1. Now we know that if A1 is 1, then A2 will display as 2. Similarly, if A1 is 365, then A2 will be 366. The formula is not displayed; only the value is displayed.

---

[16] One early reader complained that this example is complicated and boring; others have found it helpful in   making sense of time travel and anomalies. If the reader already has a firm grasp on the various anomalies and of the concept of actions and events moving against a static timeline, it is not essential to read this clarification.

We will do the same thing in A3, making it dependent on A2; and A4 will be dependent on A3, and on down the column, until we reach A100, which will be dependent on A99.

Now we'll come back to the top and enter a formula in B1 that is dependent on A100, that is, B1=A100-100, or something like that.

A moment ago we said that the value of A1 didn't matter; there is a sense in which that's still true. However, whatever you've imagined that value to be, now imagine that it changed. That is, let us suppose that the value of A1 was 1; erase that and type 2. What happens? With a good modern spreadsheet program and a decent computer, the value of B1 also changes. That is, if we suppose every cell from A2 through A100 is adding one to the previous value, such that A2 equals (A1+1=)2 and A100 equals (A99+1=)100, and B1 equals A100-100, then when A1 showed 1, B1 showed 0; but when A1 showed 2, B1 showed 1.

It's important to note that this change was, in strictly theoretical terms, instantaneous. The instant the value of A1 changed, the value of B1 changed. It doesn't matter if you've got a slow computer or a slow program and you didn't see the change immediately; it doesn't matter if you're aware that there's an imperceptible delay for processing time; it doesn't matter if you've used very complicated formulae which slow the machine. At the instant A1 changed, B1 changed. B1 was no longer equal to 0; it was instantly equal to 1.

Yet it is equally important to note something else. We could say that B1 changed because A1 changed; but there is a stricter sense in which that is not true. B1 changed

75

because A100 changed; A100 changed because A99 changed; A99 changed because A98 changed. Every step in the chain had to be altered before B1 could be altered. In a sense, it is absolutely true that B1 changed after A100 changed; yet that is not a temporal sense, it is merely a causal sense. When A1 changed, every item after A1 also changed. They changed in sequence, one after another; yet they did so instantly, without passage of time.

Time is like that, or at least it is so understood by many. If you were to change an event at point A1, it would immediately change an event at point B1. All those intervening moments would have been instantly altered. Yet they would have been altered in sequence, and it is a sequence we can only discover by moving through time.

Now, imagine that your fantasy computer is capable of something of which no computer is capable: imagine that you could make A1 dependent on B1 without causing an error. That is what time travel is like. Normally each moment in history is dependent for its "value" on the moment that preceded it in time. When someone, or something, travels from the future to the past, suddenly that moment's "value" is dependent on the "value" of a future event, one which is itself dependent on the "value" of the past event. Every time the "value" of A1 changes, the "value" of B1 must compensate; and every time the "value" of B1 compensates, the "value" of A1 adjusts.

Our three anomalies can each be illustrated from this.

In the example we've just described, after all the events presented, B1 always winds up equal to A1-1. If A1 is 1, B1 is 0. What happens if we make our formula in A1, A1=B1?

It should be evident that if A starts at 0, B1 will come to -1; but at that moment, A1 will change to -1, causing B1 to drop to -2, causing A1 to go to -2, B1 to -3, and spiraling downward, repeating the loop perpetually forever without ever repeating any of the data. Also, not only do these two numbers change, all the steps between them are undergoing similar changes. The entire line we've created cannot stabilize.

Let's expand our spreadsheet a bit. Let us suppose that B1 is the beginning of a new decade. B2 is the next event. Just as the value of A2 is dependent on the value of A1, so the value of B2 is dependent on the value of B1. What is the value of B1? Its value is in flux; it is constantly changing. B2 cannot have a value, because it cannot derive a value from B1. This is different from A1, which is situated in the chain such that, from its perspective, the value of B1 is momentarily stable. B1 only changes after (sequentially) A1 changes. B2 cannot proceed from B1 because it's outside the loop.

If you have not recognized it, this is the Sawtooth Snap. In this pristine mathematical environ, it is perpetual, never resolving to either N-jump or Infinity Loop. Although this is the simplest to create on a spreadsheet, it's a lot more complicated in reality. In reality, we're not only changing the values—sometimes we're also changing the formulae.

The next example is a bit tougher to set up. Let us suppose, however, that the starting value of A1 is 1 (as it was in our original example). Let us also suppose that through the wonders of math, the outcome of our loop is B1= -A1, that is, if A1 is one, B1 is negative one. Now, let A1 equal B1. What happens?

A1 started at 1, and resulted in B1 being -1. That, however, changed A1 to -1. Our contracted formula tells us that B1 must equal -(-1), which is of course equal to 1. Now A1 is back to 1, B1 to -1; A1 becomes -1 and B1 becomes 1.

We have not constructed the intermediate numbers this time; however, it's clear that however they're derived, they, too, are constantly changing. However, they're not changing in the same perpetually different manner as they were a moment ago. They are merely switching between two values, that which springs from A1=1 and that which springs from A1= -1.

It is also clear that once again our continuation in B2 is impossible, because we have no fixed value for B1.

This is of course an Infinity Loop. Although a bit tricky to create in the example, it proves to be the easiest to create in reality. The most common way it is created is that the value of A1 is alternately made dependent on or independent of B1 (that is, the time traveler alternately does and does not make that trip); that's beyond the ability of the illustration, but clearly a hazard of reality.

Next let us suppose that we're back to our original formula, where A1 starts at 1 and B1 winds up being equal to A1-1. Now, let us insert in A1 the formula A1=B1+1. What is this result?

For the mathematically inclined, let me quickly suggest that if A1=B1+1 and B1=A1-1, then A1=(A1-1)+1; solve for A1 and we get A1=A1.

For those for whom that was more confusing than enlightening, what we've just done is create a string of formulae which result in A1 not changing. Whatever the initial value inserted in A1, when it is replaced by the formula it retains that value. Since it retains the value, all of the equations dependent on it also retain their values, including B1.

Since B1 is no longer changing to keep pace with A1, B2 has a stable value from which to proceed.

This illustrates the N-jump. Although the cause of the value at A1 has changed, the value itself has remained the same, and all the values springing from it are likewise preserved. Although this did not require much effort on our parts in the spreadsheet, it is a very difficult object to achieve in reality—and the only one truly desirable.

Before you reject this because of fault protections and overflow errors, let me again clarify that I understand these examples won't work on our computers. Those of you who understand computer spreadsheet programs must accept that this is some sort of fantasy program that runs properly despite those errors.

It hopefully is clear that the Replacement Theory of time is not dependent on an idea of time being in motion. Time can easily be perceived as the playing field against which these events occur. Events still occur in causal sequence, and it can be said that causally the effects are instantaneous. That does not alter the fact that we normally will encounter the changes temporally, because we have to arrive at Noon on Tuesday—or at Cell A25—before we know the value there. On the other hand, were we to leap forward to Noon on Tuesday or Cell A25, we would find that the value has already been established based on changes made in the chain before it.

It is also hoped that seeing these three anomalies in this context will help some to understand how and why they do what they do.

It should also be explained again that this is an illustration; it is not the theory. The theory was developed and expounded for quite a number of years before this particular illustration was devised. Don't fall into the error of thinking that because this illustration shows so much so well that it is the source of the ideas. It is a relatively late way of presenting the ideas.

The Butterfly Effect

In discussions about time travel theory mention is often made of the butterfly effect. There is a movie named for it (obviously, *The Butterfly Effect*), and it is referenced in other films. It is technically not part of time travel theory, but part of chaos theory, from the event which launched the concept. It surprisingly has nothing to do with a time traveler killing a prehistoric butterfly as happened in Ray Bradbury's *A Sound of Thunder*—a story which predates the introduction of the theory by eight years—but instead traces to the work of meteorologist Edward Lorenz.

In 1960, Lorenz was working with a program in twelve equations which modeled weather. His program ran to six decimal places; but in wanting to repeat a segment of it from the middle, he re-entered the data to three decimal places. The results were so completely different that he had to investigate the reason. Prior to this, values beyond the third decimal place were considered insignificant—no one could measure to that level of accuracy, and everyone assumed that fractions smaller than that were of no consequence. Yet the very slight rounding that had occurred by dropping those digits changed everything.

The point was not lost on the scientist: very tiny incremental differences in the values of data accumulate to huge differences in results. Eventually someone (possibly Ian Stewart) suggested the illustration that the movement of the wings of a single butterfly would disrupt atmospheric conditions sufficiently to make the difference between the existence or non-existence of a tornado elsewhere in the world a month later.

It should be understood that chaos theory does not claim that anything which happens is random; it retains the foundation in physics that all effects are the consequential results of causes, that everything which happens is part of causal chains. Where it breaks is in asserting that for much of reality we cannot know all the factors involved in the causal chains. We know that if we hit the cue ball precisely right, it will transfer its energy in a predictable way to the ball it strikes, sending that ball on a predictable course. We cannot account, though, for imperfections in the surface of the table, changes in the air currents of the room, vibrations coming from the floor, slippage of the tip of the cue against the ball, or interference of residual chalk to the impact—all of which can alter the outcome. In so closed a system as the pool table, those variables ordinarily do not matter (barring, for example, an earthquake or sudden gale); in an open system such as the world, they become critical.

This matters to time travel because, barring fixed time theory, a time traveler cannot avoid changing the past—he absorbs and reflects photons, alters temperatures and air currents, introduces vibrations, and even with the best precautions (as in *A Sound of Thunder*) makes at least some miniscule changes to the past. Those miniscule changes are in theory amplified—much as the genetic problem discussed below, each change cascades into more changes until the probability of not altering something critical becomes negligible.

This becomes particularly problematic for Replacement Theory: if indeed the presence of a time traveler in the past has an effect amplified over time, the fact that he ate the particular hamburger someone else would have eaten or took a breath of air creating an altered airflow or simply stood in one place changing airflow and mass for a moment, could have unforeseeable consequences of devastating proportions; and the further into the past the change is initiated, the greater the potential degree of change in the future. Replacement theory analyses generally ignore this factor: not every flap of a butterfly's wing creates or prevents a tornado, and an analysis can only cover what would happen based on the known causal chains, not what might happen from the changes no one could have noticed. Yet as the Sword of Damocles it hangs above every trip into history, with the certainty that something will be different in the future that could not have been predicted.

The Genetic Problem

In *Kate and Leopold*, Kate leaves her late twentieth century advertising executive job to marry Duke Leopold in the late nineteenth century; Stuart is their descendant, proving they had children. Yet before Kate can marry Leopold she must be born, and thus there is a history in which Leopold does not marry Kate. Yet we know that he will marry someone; in fact, we know that he would otherwise marry Miss Tree, of the Schenectady Trees (the

"Miss Tree" girl). In that history, Stuart will never be born, and it is likely that there will be other descendants instead. When Kate marries Leopold, those other people, the descendants of Leopold by Miss Tree, are never born. Yet the now eligible desirable high-society Miss Tree will certainly marry someone else, and that fortunate fellow will not wed whom he would have—whom he did when she married the Duke. The displaced bride in turn finds another groom, and in a ripple that passes through New York State's upper class scores of matches are shuffled, resulting a century later in perhaps thousands of people never born, thousands born instead.

This is the genetic problem: anyone from the future marrying someone in the past creates and probably uncreates innumerable lives. It is not absolutely necessary that the time traveler enters a relationship. Apart from Marty's specific efforts in *Back to the Future*, Lorraine would have missed connecting with George McFly and ultimately would have married someone else, possibly Biff. However, Marty created quite a stir; how many other girls may have ignored boys who simply were not him? The reverse effect also occurs: when in *Millennium* Louise brings Smith home from the past, are there women whose hopes of marrying him are dashed, who settle for another suitor sooner instead of snubbing a few more men in the hope that Smith's eye will fall on them? Is it possible that some years later Smith might have married someone else had he not been removed from history? In the flow of history, who is born is one of the most important factors in forming the future, and thus these relationships and the shifts they trigger are quite significant. The addition or subtraction of a single individual in any generation can ripple through the population in a very short time.

This is also why deaths matter. In the original *Star Trek* episode *Tomorrow is Yesterday*, Spock at first concludes that because there is no mention of Captain John Christopher in the historic databanks it is not necessary to return the accidental passenger to Earth, but then discovers that Christopher's as yet unborn son, Colonel Shaun Geoffrey Christopher, would lead the first Earth-Saturn mission. This is a dramatic example, but the fact that Captain Christopher might have had additional children is itself sufficient to make him vital to the preservation of history, and even the fact that any given individual might be perceived as a desirable mate by someone has the potential to change the future population of the world. Captain Christopher could have been Kirk's great-great-great-great-greater-than-that-grandfather—or indeed Spock's or McCoy's, and given the number of people aboard the Enterprise the odds are good that at least one of them would be impacted by the life or disappearance of this one person two hundred years in the past. You cannot make changes to the population with impunity; the consequences are too complicated to predict. Every individual matters, because each individual might influence who is born in the next generation, and one change in that will likely have more impact on the world as we know it than stepping on a butterfly in the Jurassic age.

Examples abound, notably in *Timeline*, in *H. G. Wells' The Time Machine* (the 1960 version), *Men in Black III*, and many others. It is a common problem and difficult to avoid—even in stories in which it is not evident, there is often the potential that the presence of the time traveler would disrupt a relationship, a butterfly effect type problem that cannot be prevented. It may be the most dangerous aspect of time travel.

Related to this is the economic problem. When Montgomery Scott gives the formula for transparent aluminum to Doctor Nichols (in *Star Trek IV: The Voyage Home*), he comments that once the man figures it out he would be "rich beyond the dreams of avarice". Whence, though, does that money come? Someone else would have invented this and become rich from it, and that person now is not rich unless he invented something else, which would have been invented by someone else. This shift in the flow of capital will impact the success and failure of corporations, the rise and fall of politicians and governments, the general distribution of wealth, and rather directly the genetic problem—Doctor Nichols' children will undoubtedly attend different schools, meet different people, and marry spouses they would never have met otherwise. This shift in the flow of money changes history, and once again the effects can be unpredictable.[17]

---

[17] As to the impact of this in the *Star Trek* universe, the core plotline includes a major societal and economic collapse in the twenty-first century, from which the future utopia of the Federation arises before the first episodes of the show, and thus many problems created by tampering with the past can be overlooked as having been cancelled by the disaster. Maybe.

Analyzing Examples

To this point everything has been rather theoretical. There have been examples offered, but only briefly. As a surgeon better understands anatomy and physiology after dissecting a few bodies, so it will help illumine the theories of time if we take our scalpels to some time travel examples. However, as one of my early fans likes to remind me, we have no examples of actual time travel.

What we do have, though, are time travel stories.

There are some books, many of them excellent. However, books being what they are it is unlikely that most readers will have read any particular book selected for analysis. By comparison, movies are accessible. In our age, someone who wants to watch a specific movie can find it, usually at a relatively cheap cost, and can watch it in a couple hours. Thus movies have been the go-to examples for time travel expositions since the launching of the Temporal Anomalies web site, and will provide our examples here.

In examining a movie, it is important to decide what it thinks is happening as it portrays its events. The better time travel films adopt a theory of time, usually more or less like one of those already discussed. It then becomes the task of the analyst to determine whether the film is consistent with its own theory, and whether the theory adequately resolves all the temporal issues within the story. Beyond that, there is the question of whether the story is resolved as well or better by alternative theories.

To provide multiple examples, four movies have been selected for various reasons.

The first film presented here is *Back to the Future*, what is called *Part I* to distinguish it from its two sequels. This film was chosen partly because it was one of the original movies analyzed—it was the second analysis page posted—but also because it makes very few errors, being mostly consistent with its own theory of time. The sequels are considerably more problematic, but the analyses of those still on the web site are adequate and would add little here.

The second of these is *The Terminator*, first movie in the series. It is a classic at this point, and still much debated, and the more so as its sequels added to it. It is also the first movie, along with its first sequel, which was analyzed and posted to the original Temporal Anomalies web site. That rather dated analysis was somewhat updated in a later post when the fourth film, *Terminator Salvation*, required fully understanding the history of the world to that point, but in both cases the presentations were relatively brief and tried to maintain consistency with subsequent films in the series as much as possible. Hopefully here we will provide a fuller exposition.

Third up we have *Los Cronocrimines*, also known as *TimeCrimes*, a Spanish-language film available with English subtitles and dubbing which provides an intriguingly intricate interlacing of a time traveler's interactions with himself. Although there have been few changes from the original analysis for this publication, it is a lesser-known film worth including.

The final film also focuses on the time traveler's interaction with himself, in a much more complicated way, as *Predestination* gives us predestination paradoxes laced into each other. When asked what movies should be included, this was a fan favorite, and so is here by popular demand; still, it is the kind of nightmare of temporal insanity that time travel fans love, and worth inclusion.

Several other films were considered. *The Twelve Monkeys* is a very popular film, but its complexity, not to mention the sheer number of time travelers making multiple interacting trips, is enough to warrant its own short book (which might become a reality if this one does well enough). The delightful *11 Minutes Ago* was originally on the list for its well-crafted temporal interlacing and plausible resolutions. Other nominees included *Primer*, *Donnie Darko*, and *Bill & Ted's Excellent Adventure*, but while these movies have value they aren't particularly good examples of the application of any theory of time.

Analyzing Back to the Future

When the Temporal Anomalies web site launched in 1997, taking the time travel theory sketched in *Multiverser: The Game: Referee's Rules* and applying it to popular time travel movies, the second analysis posted was for the movie *Back to the Future*, later to be referenced as (*Part I*) to distinguish it from its sequels.[18] Although since then understanding of the theory has expanded greatly, the original analysis has stood the test of time, because it was a relatively simple film as far as the time travel is concerned. It is also one of the best illustrations of Replacement Theory, as it gives us the framework of an original history, then shows how it is altered.

That means, up front, that we can rule out Fixed Time Theory as a foundation for the film: history is changed, and therefore under the rules of the story history can be changed.

In this story we see very clearly the complications of altering a timeline. As Marty McFly explores the world of 1955, he is confronted with the fact that the history he learned of the events of his parent's life is changing before his eyes, and his own existence is in jeopardy. He has interfered with the meeting of his parents, and must correct the situation before it's too late, or he will cease to exist. This is very valuable to us, because right from the beginning we can see two distinct timelines: the original

---

[18] The first analysis posted to the site covered both of the first two *Terminator* movies as if they were a single story. Although the three *Back to the Future* movies were presented as a single continuing story (and in fact the second and third were filmed as if they were one movie in two parts), there were enough differences between them to make it more practical to address them individually, and so the three parts were posted in three separate pages.

A-B timeline in which George McFly gets hit by a car, marries the daughter of the man who hit him, and lives the rest of his life as something of a nerd and a loser; and the altered timeline, in which Marty has prevented that, and is trying to correct it so that George will marry Lorraine and Marty will be born.

The Beginning

In our A-B timeline, shortly after point A, George McFly meets Lorraine Baines (through the aforementioned accident), marries her, and gets a job working for the bully who has terrorized him all his life, Biff Tannen. They have three children, the third of whom takes an interest in music, and so connects with Doctor Emmet "Doc" Brown, who provides him with access to technological equipment he might not otherwise have had. As a guitarist myself, I am quite aware that instruments, amplifiers, effects boxes, mixers, microphones, and other equipment all cost a great deal of money. Being able to repair or even build some of these things for your own use is a bonus, and Doc has the skills to do so. Marty would find his friendship with Doc useful in this way, and Doc would enjoy having the lad around, teach him bits about the equipment, and use his help in some of his experiments. This is the relationship we see at the beginning of the movie.

Doc Brown has an important history, but the significant part occurred in 1955 when he hit his head and awoke with a vision of something he would ultimately call the flux capacitor. What he thought it would do at that point is unclear, but he knew it was important. It ultimately becomes the basis of a time machine, thirty years later.

It matters that Marty has two older siblings, a brother Dave who works at McDonalds and a sister Linda who is similarly failed in life. His father George McFly has done nothing but watch television and work for a boss who takes advantage of him. His mother Lorraine Baines McFly is overweight and largely ignored by her husband; she has rather strict almost Puritanical views about the conduct of young people, which she insists is nothing like what she did as a teen and completely inappropriate.

Marty is the only member of the family showing potential. He has his own band, which is actually pretty good, in which he plays guitar and sings. He has a cute girlfriend, Jennifer, who is very encouraging and believes that he can do something with his music. Vice Principal Strickland thinks he is a loser who will always be a loser, just as, in Strickland's opinion, his father was a loser; but he has a chance.

Doc Brown has spent thirty years and a fortune building a time machine from a flux capacitor and a Delorean. He believes it will take him to the future. Unfortunately, it requires 1.21 gigawatts to power the time travel portion, which he generates with doses of plutonium he stole from some Libyan terrorists who believed he was going to make a nuclear bomb for them. They are after him.

On October 25th, 1985, shortly after midnight, Marty joins Doc at what is called the Twin Pines Mall—a significant name in the time travel story. Doc has Marty video what he expects will be the revelation of his discovery of time travel, but they are interrupted by the arrival of the Libyans. Doc is shot; Marty leaps into the Delorean to attempt to escape, and in fact does so in an unexpected way. At 1:21 in the morning he hits 88 miles per hour, the time circuits respond, and he travels back thirty years, to a time before his own birth, to a time when his parents were at the beginning of their senior year in high school.

Changing History

Marty arrives in 1955, badly disoriented. He crashes into a barn that wasn't there a moment ago in his reality, gets chased by the farmer whom Doc had told him once owned all that land, and in his escape runs over one of two young pine trees the farmer had hoped to use to begin a pine tree farm.

It is important to recognize that at this point Marty has already changed his own past—not, perhaps, in a way most would notice, but when ultimately he returns to the future he races to find Doc at the Lone Pine Mall—the Twin Pines Mall no longer existing because he killed one of the two young pine trees for which it had been named.

This should also alert us to the fact that this is not his history. He is making changes which might have unpredictable effects. The farmer, his wife, and their two children have seen what they believed was a spaceship and an alien; how will that impact their futures? From there he begins interacting with other people, giving Goldie Wilson the idea that he could be the mayor in the future, trying to order a non-existent diet soda at a malt shop, meeting and following George McFly, preventing George from being hit by Mister Baines' car and so meeting Lorraine, and becoming entangled with Lorraine.

This moment of his arrival in 1955 is point C, the exact same moment in history as point A, but now different because Marty McFly is there; had he made no other changes in time, his presence alone would be enough to make this a different timeline. As this timeline develops, Marty prevents the initial meeting of his parents, and, aided by Doc, recognizes that his existence is threatened—but not only his. If Marty McFly is never born, all of time is trapped in an infinity loop.

Follow this: let us suppose that Marty is unable to repair the damage he has done; let us further suppose that no one else intervenes to repair this damage. (It would be possible that somehow George and Lorraine would still meet and marry, perhaps after high school, under other circumstances, and Marty would still be born; as unlikely as it seems, it is important to remember that Marty does not have to be the one to fix the timeline, as long as the timeline is repaired.) Since George and Lorraine do not meet, George does not marry Lorraine, and Marty is not born. Since Marty is not born, he does not return in Doc's time machine; therefore he does not interfere with the meeting of his parents, and therefore he is born, and therefore he does. These two alternate histories would repeat in perpetuity, an infinity loop.

However, that is not what happened. Marty did arrange for his parents to get together (albeit not quite as he had planned), and the incidental result was that all of the future was changed. We get a glimpse at that altered history at the end of the movie.

Quibbles

I will interrupt this retelling to make some minor complaints.

Very shortly after his arrival, Marty encounters Goldie Wilson, working in the malt shop. Goldie is talking about how he isn't always going to be a malt shop employee, but is going to be somebody important someday, and Marty says, "That's right, he's going to be mayor." Immediately Goldie embraces this idea, and tells his boss that yes, one day he will be mayor, despite the fact that he is, as his boss calls him, "colored".

This little throwaway is not a predestination paradox—it's just made to look like one. We know two significant truths about the original history. One is that Marty McFly never arrived in the past and so never said those words to Goldie Wilson. The other is that Goldie Wilson is the mayor. So in the altered history Marty might have been the first person to give Wilson the idea of becoming mayor, but the upwardly motivated Wilson obviously came upon the idea some other way when Marty was not there. That also means that Marty does not have to make that statement in the next iteration of history (that's coming), because whether or not he mentions it, Goldie Wilson will eventually decide to run for mayor, and will win.

There is the problem of the fading photograph. Early in the story Marty produces a picture of himself with his two siblings; Doc notices that the top of Dave's head is missing. Gradually Dave disappears from the top down. Yet this is nonsense. The Marty McFly who carries the photo comes from an original timeline in which those people all exist; he was born, he has his own life history and existence. These things will continue to exist as long as he does. Even if our replacement timeline approach is not the truth, a logical consideration of the matter will quickly dispense with any notion of this "symptom" of the temporal anomaly being possible. Marty McFly is in theory vanishing because his parents have not gotten together in the future which is his past; but it is not yet determined whether they will or will not get together. The fact that he is attempting to restore the future history proves that they might still get married. His existence is not a percent probability—he is not less real or more real based on the likelihood of his birth! Were that the case, he would become non corporeal, ghostly, and lose the ability to affect reality at all. At the moment he arrives in the past, he either is real or is not real, and he remains in that state until he either leaves the past or dies there. The picture either exists or it does not—the changing images are foolish. How could the theoretical other picture even exist? No version of Marty McFly would be carrying a picture of that place with no one in it; nor would the picture make any sense if he were standing on one end, alone in the picture. And a picture of Marty's brother without his head is far more absurd! No, this Marty has a picture, and the picture does not change.

Similarly, Marty is who he is. He may have no future, but he has a past in the original timeline. He doesn't slowly forget how to play the guitar; he doesn't start to fade out of existence. Logically, if those things were true, then Marty would have vanished completely the moment he prevented his father from being hit by his grandfather's car—but then if that were true, he would cease to have been in the past at all, and would not have saved his father, and we would have that flicker loop previously suggested. No, Marty remains real and intact until his existence is completely undone. Either he arrived in the past and saved his father, or he didn't.

For the sake of the film, the fading photo and the fading Marty make good plot devices, communicating quickly to the viewer the serious nature of the situation. Marty McFly might cease to exist, and that is serious—not merely for McFly, but for all of time.

Another Change

At the end of the movie we see the impact Marty had on his own history. George McFly is a successful science fiction author (undoubtedly encouraged by his son years before), and the family is much more upscale. Dave works in an office, Linda is popular, and their mother is thin, athletic, and open-minded, their father flirting with her constantly. Biff is not George's boss, but an auto mechanic who is grateful for the opportunity to clean George's cars for a little extra cash. It is a fortunate point that no drastic changes occurred—the family did not buy a different house, but made this one nicer; they had the same three children, Jennifer was still his girlfriend, and he even was planning the same weekend in the mountains—but the changes are major, and Marty's life was very different.

Yet we need to understand something that the film misses.

Marty returns to the future ten minutes before he left for the past, and he manages to watch himself leave. Who, though, does he see leaving? The kid looks like him, dresses like him, and does exactly what he did—but it is not him. When the dust settles Marty will go home and in the morning awaken to find the affluent McFly family. Yet that house didn't just change overnight. It changed over thirty years. The Marty that he just saw leave grew up in that house, with that family.

Marty McFly grew up a different person—not drastically different, but slightly different. His family now has money; he doesn't necessarily need Doc to help him put equipment together. However, in this timeline, Doc knows that Marty will go back into the past, and so Doc has an interest in preserving the timeline. Remember, he's the one who best understands the temporal problems; given those 30 years, he would have worked out the necessity of Marty McFly returning to the past, interfering with his parent's meeting, and bringing them back together, without which we are caught in a different infinity loop, as those events keep happening and "unhappening" with each cycle. Thus Doc would cultivate the relationship with Marty, hiding the truth about the past from him, so that Marty would return. Doc has also read Marty's note, and recognized that in order for him to save his own life, 1) he must take some steps to prevent the Libyans from killing him, and 2) Marty must believe that he was shot. The bullet-proof vest is the best answer. Marty—the more affluent Marty of the C-D timeline—sees Doc shot and escapes to the past, not knowing that Doc was wearing the vest. Still concerned for his friend's life, Marty still writes the note, so Doc is still saved.

But the Marty who returns from point D reaches point E—not point C nor point A—because his information is completely different. In his history, George's friend Marty was trying to get George together with Lorraine, and Biff got in the middle, so George K.O.'d Biff, impressing everyone, especially Lorraine, who danced with him, kissed him, and married him—living a very different happily ever after than the one in the A-B timeline. This Marty, the one born and raised in the C-D timeline, knows nothing of George being hit by a car. He (and Doc) have a lot more extrapolation to do in order to realize that *Marty is the Marty for whom he is named*—the friend whose influence helped bring them together, the friend who encouraged George to share his stories and become an author, and then disappeared after The Enchantment Under the Sea Dance, never to be seen again. And so the E-F timeline is in some ways different from the C-D timeline which we saw. We have a sawtooth snap; history repeats itself until either it falls into an infinity loop (by a change drastic enough that Marty cannot return to the past or fails to unite his parents) or advances to an N-Jump termination (in which Marty's next return to the past will repeat the previous one in every detail).

Clarifying this, if Marty on this E-F timeline fails to bring his parents together, then at F he no longer exists, and we return to the A-B timeline, forever trapped in a cycling causality. If he fails to interfere with his parents' original meeting, then at F we have restored most of the circumstances which existed at B, and the other Marty will make the time trip to set up the C-D timeline. If he succeeds in bringing about the same results as the C-D timeline, then history continues as at the end of the movie—except of course that the Marty who wakes up in the upscale McFly home is not at all surprised, because things are as they were when he left.

Thus the Marty McFly whom we see at the end of this film does not exist. Although he returned before he left, at the moment that his other self leaves for the past, that timeline is erased. The only future which exists is the one at the end of the E-F timeline. Remember: for the Marty McFly who is seen leaving by the Marty McFly who has just returned, that more affluent existence, that altered history we just watched, is his. He is not the same McFly. And when he returns, he will reach a time moments before point F, the end of the timeline he is on his way to create.

That timeline will be different, but it is possible that the end will be much the same (without the surprised Marty). Doc might have his bullet proof vest (a gamble, as he did not know that they would not shoot him in the head or use armor piercing bullets), and so might be alive, if somewhat bruised. The McFly family might be the affluent version. And since it might be so, and the story continues that it is so, it is reasonable to conclude that on the last timeline of the sawtooth snap—whether it is E-F or Y-Z—time continues into the affluent future. *Back to the Future*, in its first part, allows the future to continue.

Don't misunderstand. The point is that it is possible for this to go right. It is also possible for it to go wrong in uncounted ways.

We know that Marty interfered with his parents' meeting. He fixed that. However, he created quite a stir by his presence, with many other girls asking who he was. He may well have damaged some other relationship—a couple who never got together because she was too interested in Marty to notice the boy she would have married, or the couple who broke up because he was offended by her attention to this new kid. That means we have the genetic problem: this couple did not get together, so each of them found someone else, who then did not connect with the person they would have in the unaltered history, and we have a couple shuffle, changing the births of an unknown number of children in Marty's generation. It may be that one or more members of Marty's band were never born; it might be that Jennifer Parker was never born. This could have happened—but it might not have happened, and so we can say that although the resolution in the film is not guaranteed, it is possible.

Meanwhile, the fact that Marty managed to reconnect his parents does not necessarily mean that he saved himself. The relationship between George McFly and Lorraine Baines McFly is very different from the one that produced Dave, Linda, and Marty. They might have had more children, or had them sooner. As we see in *About Time*, those kinds of changes can impact the genetics of a single child even if everything else is unchanged—George and

Lorraine might have been the parents of Sandra, Dorothy, Michael, and Jane, all of whom were born before 1967 (the year of Marty's birth) to this more amorous couple. The odds are actually against Marty being born in this new history of his parents' relationship—not impossible, just improbable. However, again it is possible.

Doc's life has also been changed. He knows that that thing he envisioned, the flux capacitor, is what makes time travel possible, and he has actually seen the time machine and knows its power requirements. In the original 1955, he probably had no idea what he had found, and he spent decades experimenting, trying to figure it out. In that time he still tinkered, still tried to invent things. Most of those probably failed—but it is possible that something he created was integral to making the Delorean do what it ultimately did. His creative efforts have turned in a different direction, becoming more focused on building the device he knows will work. In the process he might miss a necessary step. He won't have learned the same things over that time. Still, what matters is that he produce the time machine and cultivate his relationship with Marty, and these outcomes are possible. They are actually more probable than that Marty would be born, and we can accept that they are likely to have happened given the birth of Marty.

There is also the less important question of Marty's relationship with Jennifer. After all, the Marty who grew up as the son of a successful science fiction author (and George must be selling short stories to magazines and anthologies to have the lifestyle they have) is going to have had a different upbringing from the one whose father was a low level office worker bullied by his boss. His mother's attitudes toward his relationships are also significantly different, and given the changes in his parents and their relationship we can be fairly certain that the affluent Marty is different in more ways than that his family has some money. We don't know what attracts Jennifer to Marty, or Marty to Jennifer, but if Marty is not the same person, there is a good possibility that he has a different girlfriend. Even when people are nearly the same we have this—in *The Magicians* Penny died, but was replaced by a version of himself from a parallel universe. The Penny who died was boyfriend to Kady, but the Penny who replaced him did not even know Kady and was instead boyfriend to Julia. It doesn't take much to interfere with relationships. Yet it is still possible that Marty and Jennifer would connect, even given that this is a different Marty. It is even possible that they would be planning a date at the lake that weekend, and that he had wanted that pickup truck, and although the probability of these being so drops precipitously, they are still possible.

We thus recognize that it is possible that after a brief sawtooth snap (to accommodate the change from the original Marty to his affluent version) the story resolves to an N-jump.

An Alternate Explanation

To this point we have been considering the film primarily as an example of Replacement Theory, because it appears to us that Marty McFly has traveled to his own past, changed that past, and returned to the altered future. The question, though, is whether there is another possible explanation for the events we see in the film.

It is patently obvious that this is not a Fixed Time Theory story. The core of the story is the danger that Marty has already changed his own history and needs to change it back, and in the end he arrives in an altered future. Fixed time will not work here.

What, though, of Multiple Dimension Theory? In *Back to the Future Part II* Doc attempts to explain what has happened to them by using a Divergent Dimension Theory model. Indeed, the core of the story could work as a Divergent Dimension Theory story: Marty travels from 1985 to 1955 and creates a new and different history of the universe. When he travels forward he is in the universe he created. He does not have a problem with his divergent doppelganger, because that version of himself also travels to the past, and he is there in time to see that happen. He is not the Marty these people know, but since the other has departed they might wonder why he has changed but would never guess the cause. There are many aspects of this world and its people which will be unfamiliar to him, but we already see that unfolding.

It has a hidden problem: the version of himself who grew up in this world has just left for 1955. In theory, he will land in his own history—but in his history there is already a divergent Marty in that field driving a duplicate Delorean in what is likely to be exactly the same place. How that works is probably disastrous, but assuming that it doesn't kill both of them and they can't merge into one person and one vehicle,[19] this creates an entirely new set of problems for them. However, we can't say that this is not what happened, as we don't see that version of events.

However, the film tells us that this is not its theory of time. The value of Divergent Dimension Theory is that you are not changing your own past. Certainly it can be argued

---

[19] There is further discussion about interactions between temporal duplicates later in the text.

that when Marty returns to the future he lands in the past of the other Marty, a world he created. However, even though we discounted the fading photograph and the fading Marty as plot devices which would not really happen, they only make sense if they are telling us that Marty is erasing his own past, undoing his own existence. That would not happen under Divergent Dimension Theory, as Marty's past would still exist in the other universe.

That problem applies equally to Parallel Dimension Theory: if he is changing the history of another universe, he is not undoing his own existence. Only some version of Replacement Theory can explain the first *Back to the Future* film.

The sequels are more complicated, but not part of this analysis.

Analyzing Terminator

The first analysis to appear on the Temporal Anomalies web site covered the movies *The Terminator* and *Terminator 2: Judgement Day*. It was a bit rough, partly because it treated the two films as one story, but it was later refined when *Terminator Salvation* was released and a general reconsideration of the entire series to that point was published.[20] However, it was still done as if all the movies formed a single story—which at that point still made sense, but started to fall apart as the franchise continued. Thus it makes sense to include an analysis of just the first film, both because it could use some fresh insights and because it is still an excellent example of time travel.

What makes the film so challenging is it is built on two interlocking predestination paradoxes, both of which escalate as sawtooth snaps, feeding themselves and each other.

To put the events in chronological sequence for the timeline we see, limited to the content of this first film, it begins with the arrival in 1984 of a killing machine from the future, a T-800 Model 101 Terminator. It has come with the programmed mission that it should kill everyone in the Los Angeles area named Sarah Connor, because its sender, a future artificial intelligence called Skynet, wants to prevent the birth of someone named John Connor, whose mother was Sarah. Minutes after that, a person named Kyle Reese arrives from the future, having been sent to protect Sarah and ensure the birth of John, the rebel

---

[20] This was originally published at TheExaminer.com, as I was the Time Travel Movies Examiner for several years. It was then copied to the Temporal Anomalies section of mjyoung.net.

leader who is starting to win the battle against Skynet. Kyle Reese is probably already in love with the idea of Sarah Connor, and in the tensions of trying to escape the killing machine they become enamored with each other, and Kyle fathers John Connor. The chase and the battle continue, until Kyle is killed but the Terminator is destroyed.

That final fight happens in the factory of a company called Cyberdyne, and although the company finds the remains of that future robot they deny it, and then study those parts so as to create a defense computer system called Skynet, which is delivered presumably to the United States Air Force to operate its weapon systems. In 1997 the system asserts its independence and attacks humanity.

That war continues for perhaps decades, with John Connor leading the human resistance against Skynet and its increasingly capable killing machines. Then at some uncertain future moment Skynet has managed to develop a time travel projector, and sends one of its most sophisticated killing machines back to 1984, with instructions to kill Sarah Connor. A few minutes later the resistance captures the time machine, and John Connor sends Kyle Reese back to protect Sarah Connor.

That is everything we know with certainty from the first movie.

A Fixed Time Solution

Many fans of the movie assert that it is a perfect Fixed Time Theory story, that everything in it happens in a self-supporting loop that simply always is.

Arguably the second movie makes a fixed time solution impossible: Sarah, John, and the T-800 working together destroy Cyberdyne and prevent the development of Skynet. That this movie is a continuation of the story from the first is easily defended: unlike later films, this was also from James Cameron, and included Linda Hamilton as Sarah Connor, obviously intended as a sequel by the man who created the original. If his view of time travel in the sequel makes it possible for Sarah to alter the future, it is reasonable to assume that this was his view for the first film, granting him the courtesy of assuming his view of the subject was at least consistent. However, the second film could be dismissed as parole evidence, that is, information not contained in the original film and therefore not binding on any consideration of that first film. Whether the movie works under Fixed Time has to be decided based on its own content, not on anything added later.[21]

The issue, then, comes down to the two complicated interwoven paradoxes.

It seems as if the first paradox is created when Skynet sends a machine to the past to kill Sarah Connor. That machine fails in its mission, but its destruction leaves remnants of future technology in the possession of a technology company, Cyberdyne, which uses those parts to learn how to build Skynet. We thus have the loop that Skynet has caused its own birth by sending technology to the past. If Skynet never sent the T-800 to the past, Skynet would never have been created.

---

[21] For example, our rules of analysis include that outside information such as "Director's Cut" editions or interviews with the directors, writers, or cast do not dictate our understanding of what actually happens in the movie, which is based entirely on what appears on the screen in the publicly released version.

However, there is another complication. Skynet would never have sent the T-800 to the past were it not that it needed to eliminate John Connor. John Connor is only born because Kyle Reese comes to the past to protect Sarah Connor from that T-800. If the T-800 doesn't come to kill Sarah, Kyle doesn't travel to the past; if Kyle doesn't travel to the past, John is never born; if John is never born, the T-800 is not sent to the past to kill Sarah.

Occam's Razor tells us always to accept the simplest explanation of events. We are looking at self-causing events: if John Connor is not born, he is not born; if Skynet is not created, it is not created. Further, if John Connor is not born, Skynet is not created, and if Skynet is not created, John Connor is not born. Anything that only happens if it happens, anything that causes itself, does not happen. Predestination paradoxes are inherently irrational.

There are Fixed Time theorists who believe that predestination paradoxes are possible, that as long as everything that happens is caused by something else that happens, the temporal sequence of these events is not relevant. As irrational as it seems, *The Terminator* is embraced as an example of such predestination paradoxes, and it arguably does work if we accept their possibility. However, this is not the only plausible explanation.

Other Dimensions

It might be plausible that *The Terminator* works under some form of multiple dimension theory, and that should be considered. After all, it could be that there was an original history in which no time traveler arrived, then Skynet sent a machine back to kill Sarah Connor, creating a new history (or altering an existing parallel one), and then the resistance sent back Kyle Reese to protect Sarah Connor creating (or altering) another history.

There are glaring complications in this.

The first problem is where we get Skynet. Without the T-800 arriving in the past, Cyberdyne never creates that system. The logical solution to that is that in the original history someone else created a different Skynet which otherwise did the same thing—and *Terminator 3: Rise of the Machines* gives us this solution with the Autonomous Weapons Division of the Cyber Research Systems branch of the United States Air Force. It doesn't have to have been them, but in order to get an existing Skynet in an original history in which nothing comes from the future, we need an original origin of the artificially intelligent monster. It doesn't have to be the same kind of Skynet, as long as it is able to recognize John Connor as a threat and develop a time machine and a killing machine and a plan to target his mother.

The next problem is why Skynet wants to kill Sarah Connor. We have those complicated interlocking causes, that Skynet sends the T-800 to kill Sarah in order to prevent the birth of John, but John is only born because the resistance sends Kyle Reese back to protect Sarah from the T-800. Without the T-800 there is no John Connor; without John Connor, there is no T-800.

The solution to this is not much different from that for the other problem: there must be an original cause of the birth of Sarah Connor's child. For what it's worth, that child does not need to be John Connor; it only has to become a rebel leader whom Skynet wants to destroy. In our original history, Sarah was out partying. If we extrapolate that she connected with someone and got pregnant, and then her child grew up to lead the resistance, we resolve this. Notice that the T-800 wants to kill Sarah Connor, but we don't know anything about her child until Kyle Reese gives him a name—but Kyle has to come from a history in which the T-800 has already changed history.[22]

In the divergent history, the T-800 arrives in the past, Sarah connects with the guy and gets pregnant, and avoids the Terminator long enough to give birth to a child who becomes a rebel leader. Those events have to happen to create this divergent history. If Sarah never has the child, it doesn't matter (in multiple dimension theory) that Skynet never sends back an assassin—but it does mean that the resistance will never send back Kyle Reese, because there is no one to protect. Indeed, it is unlikely that Skynet will even develop the time machine, and probably won't use it, because the mission is completed.

The glaring problem is what happens if Sarah has the child. Whether or not the Skynet in this universe knows that there was a T-800 sent back by its counterpart in the original history, it has the same problem, and will resort to the same solution: it will send a killing machine back to 1984 to kill Sarah Connor. However, there is already one there—the one that by arriving created this history of the

---

[22] *The Terminator* is a complicated story. It is possible, for example, that the original rebel leader was the child of one of the other Sarah Connors, killed by the terminator before it came after our Sarah, and that our Sarah is the replacement cause of the birth of the rebel leader whose original mother was killed. This might be a more complicated solution, but it is a plausible one. However, it still requires that Sarah gets pregnant in the history in which the terminator comes to the past and Kyle Reese does not, and since nothing else changes we have to assume either that Sarah got pregnant in the original history as well, or that something the terminator did caused her to become pregnant in that first altered history.

world, the world in which a Terminator arrived in 1984 to kill Sarah Connor. There would now be two of them.

Well, not exactly. The one arriving from the second history creates the third history, and in *that* history there are two terminators. In the second history, Skynet does not know that the Terminator that arrives in 1984 is not the one it sent, and neither does the resistance, who sends Kyle Reese to protect Sarah. Unfortunately, from their perspective, Kyle never arrives in the past, because his arrival creates a new history in which he and Sarah battle the T-800. That becomes the history we see in the film—the fourth universe created, with a myriad more to come, making the mish-mash of *Terminator Genysis* entirely plausible. However, this fourth universe is a divergent from the third, and thus there must be two Terminators in it; the universe in which there is one Terminator and one Kyle Reese is nearly impossible to derive.

So Divergent Dimension Theory could not create the story we see, but instead creates a wealth of other diverging stories whose resemblance to this is passing at best.

Rewind, Replace

Our problem is that John Connor is not born unless Kyle Reese travels to the past, and Kyle doesn't travel to the past unless Skynet sends a Terminator, and Skynet doesn't send a Terminator unless John Connor is born—plus Skynet doesn't exist unless it sends a Terminator to the past.

To resolve this, we need two events. One is that Skynet has to come into existence without the benefit of having sent a Terminator to the past; the other is that Sarah Connor has to have a child without Kyle Reese. We have already addressed both of these problems in trying to establish a Divergent Dimensions solution; those solutions work better with Replacement Theory.

The first thing that must happen is Sarah Connor meets someone. One night in 1984 she gets pregnant. We know almost nothing about this child—not its name, not its father's name, not even whether it is a boy or a girl. However, this child will be important as a leader in the fight against Skynet.

The second thing that must happen is that someone creates an ultra-intelligent machine which is given the name Skynet. Our prime candidate, thanks to the third film, is the Autonomous Weapons Division of the Cyber Research Systems branch of the United States Air Force, although it could be someone else. That Skynet need not be as advanced as the one behind the scenes in the movie; it does need to be able to develop terminators and ultimately time travel, and to use our weapons systems against us.

Sarah Connor's child becomes a problem for Skynet by leading a group of resistors against it and doing so effectively. Kyle Reese is born and becomes part of that resistance.

We now have a future in which humanity, led by Sarah Connor's child, is fighting against an artificial intelligence called Skynet which uses killing machines. The only thing that still has to happen is that someone invents a time travel projector and Skynet has access to it. It decides that the best solution for its problem is to kill the mother of Sarah Connor's child before she has the child. It sends one of its killing machines back to find and kill Sarah Connor.

The date on which it does this is restricted by a couple of factors. One is that the time machine itself must be invented. The other is that Skynet apparently has to develop a killing machine encased in living tissue, because that is a limitation of the time machine. As soon as those problems are resolved, it does this. The machine leaves from the future and arrives in 1984, intent on killing Sarah Connor.[23]

One way to comprehend Replacement Theory is to think of a multitrack tape recorder, the type used in professional recording studios.[24] With such a recorder you would record a song by sending each instrument to its own "track" on the tape, allowing you to adjust the volumes of them individually to balance them for the desired sound. You can also rewind the tape to the beginning and, for example, re-record the guitar track: as the tape moves through the heads it plays all the tracks but the original guitar track, which it erases and replaces with the new version played by the guitarist as he hears the other instruments. In something analogous to that, when someone travels to the past time is rewound to the moment of arrival, and the recorder replays those parts of history that are unchanged while moment by moment events related to the time traveler are erased and replaced with the new version. Thus as the terminator arrives in Los Angeles it starts replacing the history of people in that city—some by a little, that they hear about these murders

---

[23] In the film the sense is that Skynet does not send the Terminator until moments before its doom is certain. It might be that it is concerned about the plan and does not initiate it until it has no alternative.

[24] That is, those which still record in analog on magnetic recording tape. Digital recording systems are different in significant ways.

on the news, others by a lot, that they are murdered, with various levels of change between those extremes.

What the machine fails to do, however, is prevent the birth of Sarah Connor's child. Somehow she still meets the guy, gets pregnant, and avoids the killing machine long enough to give birth to a baby who is going to grow up to be a problem for Skynet. That has to happen for the movie to make sense.

The more obvious problem is that if Sarah doesn't have the child, Skynet doesn't know that she has to be killed and so doesn't send an assassin to the past. Then she would have the child, Skynet would send the assassin, and we would have an infinity loop. If Skynet is smart enough, though, it could avoid this by having the machine shut down for a few years and then restart and deliver the information that Sarah Connor has to be killed because her child is a problem, and Skynet could act based on that information, sending the terminator to kill her.

However, there is a second problem.

History has to stabilize before time can continue beyond the departure of the assassin. The resistance can't send anyone back to protect Sarah until there is a self-supporting history in which a terminator travels to the past to kill Sarah Connor. Here's the problem: if in that self-supporting history Sarah Connor's child is never born, then the resistance does not know who she is or why she matters, and has no reason to send anyone back to protect her. That means that somehow Sarah Connor's child must be born and survive.

However, we have the other side of the problem. We think that the way it works at the future end is Skynet sends the terminator back, and realizing the danger this creates the resistance immediately sends Kyle Reese back before the terminator can kill her—but it can't work that way. Rather, once the terminator leaves the future it has arrived in the past, and whatever it is going to do in the past is, from the perspective of the future, a fait accompli, that is, already done, and the only history anyone in that future has ever known. That means that this iteration of the resistance knows—and has always known—that a terminator arrived in 1984 but failed to prevent Sarah's child from being born.

Yet for some reason the resistance decided to send Kyle back to protect Sarah.

This is a dangerous decision. After all, we know that the Connor child was born, and that means that Sarah was not killed before the birth of the child. One of the most dangerous things you can do with time travel is try to fix the past, to change history. The birth of the Connor child is the good outcome. Sending someone back to mess with that has a fair chance of preventing that birth. Yet they did it. The question is, why?

We know that terminators are relentless. This machine is not going to give up until it finds and kills Sarah Connor, or it believes it has done so, or it is destroyed. We know that Sarah managed to escape long enough to give birth to her child; we don't know how. However, we can guess that she and her child were on the run for a few years, long enough for the child to remember the mother, and then the assassin finally completed its programmed mission. It would ignore the child—that wasn't part of the program. It ultimately succeeds in killing Sarah Connor, and shuts down awaiting the rise of Skynet.

That gives Sarah Connor's child the necessary motivation to send back Kyle Reese to protect Sarah: this might save Mom.

Arriving in the past, Kyle Reese protects Sarah Connor from the assassin, but in the process prevents her from meeting her child's father. Instead, Kyle becomes the father of the child, and the child is a son, given the name John Connor.

Kyle makes several other critical changes to history at this point. Thanks to his effort, the killing machine is destroyed—but its parts are left in a factory belonging to a technology company known as Cyberdyne. They study these parts, and become the inventors of a different system of the same name, Skynet. The original inventors, and their system, are lost, part of an erased and forgotten history. Sarah learns from Kyle exactly what they are facing, and lives long enough to begin preparing her son to be the leader of the resistance in the future.

Ratcheting

One question that looms over this reconstruction is how Sarah Connor managed to escape from a T-800 Model 101 in that first altered history in which she did not have Kyle to help her. The answer is that she didn't—that is, although a killing machine was sent back in time to assassinate her, it was not *that* killing machine. Neither that original Skynet nor its robots were quite as sophisticated as the ones in the movie. Rather, a less sophisticated machine was pursuing Sarah, and she was able to outrun it for long enough to give birth to her child and find a safe home for it before it caught her. Then when Kyle Reese came back, they fought that less-sophisticated machine, and left its fragments for Cyberdyne to study.

This gave Cyberdyne a leg up in creating a superior version of Skynet, which in turn created superior terminators. Thus in the history in which the original terminator is destroyed in the Cyberdyne factory, Skynet sends back a better terminator—which means that the fight is tougher, and the parts left in the factory are more sophisticated, and Cyberdyne is able from them to build an even better Skynet. This loop will repeat multiple times before it plateaus, reaching a point at which Cyberdyne cannot learn more from the parts it gets and so creates the same Skynet.

Once that happens, we have an N-jump termination on our sawtooth snap, and time advances to the moment that Kyle is sent to the past—a different Kyle, the Kyle who knows this version of Skynet, this terminator. We get the version of history we see in the movie. Time has resolved.[25]

---

[25] It was observed in previous analyses that the birth of Kyle Reese appears to fall between the 2004 probable original Autonomous Weapons Division Skynet launch and the 1997 Cyberdyne Skynet launch, with the result that the genetic problem will mean there are two different versions of Kyle Reese. That also means that there are two different versions of John Connor. However, both John Connors will be looking for a Kyle Reese, and it is a common enough name that both will find someone of that name to train and send back to be the father. This anomaly will resolve, because the date of the launch of Skynet is not dependent on the identity of Kyle Reese or John Connor, so once that date settles so will their identities.

Analyzing Los Cronocrimines, a.k.a. TimeCrimes

Los Cronocrimines is a 2007 Spanish film released to English-speaking audiences under the name *Timecrimes*. It proves to be a fascinating multi-layered predestination paradox, but resolves to Replacement Theory with a very few assumptions. The version we viewed was subtitled and dubbed in English. An American film company had allegedly acquired the rights to do an American version of this story, which at one point was slated for release in 2012, but it has not materialized, still appearing on the Internet Movie Database[26] as "in development". It created a great deal of buzz among time travel fans when it was released. It presents as a sort of horror film; the trailers on the DVD are all for thrillers, mostly with supernatural elements. Get past the atmosphere, though, and it's a rather straightforward story. As a *time travel* story, however, it is a bit more complicated.

This analysis tells the entire story of the film in significant detail from multiple perspectives, and so it is recommended that the reader have viewed the film before reading this part of the book, so as to more fully enjoy the movie. If you have not seen the movie, you might find this chapter particularly difficult to grasp. However, it is quite good as time travel movies go, and well worth watching, so you would benefit by taking the time to view it before continuing here.

What *Timecrimes* does is create a predestination paradox, but does so in layers: the core character, Hector, is lured into traveling back in time by the version of himself who has already done so, who in turn is put in the position of

---
[26] IMDB.com

luring his self back in time and of making another trip by the version of himself who has already made that second trip. Along the way, he terrorizes the unnamed researcher who sends him to the past, and terrorizes and kills a young woman who stopped to offer him help when he was injured. Although the jacket says that the past can be changed, the story works as a Fixed Time Theory story, in which everything that the traveler from the future is going to do in the past has already been done. This is often the case with Replacement Theory stories with N-jump terminations, that the final timeline is indistinguishable from a Fixed Time Theory story.

That's fine if you can accept the uncaused cause that is essential to the predestination paradox form. If, however, you are of the opinion that anything which is its own necessary cause can never happen, then the question of whether *Timecrimes* works is a complicated study in replacement theory, looking for original causes which brought about the original situations only to be erased and replaced by the time traveler's intervention. That is not so simple in this film. Many times characters are acting with a view to attempting to make events happen the way they remember them, either as they originally happened or in a way that will appear as they originally perceived them. The problem is that many of those events make no sense at all absent that motivation. Although the final form of events is entirely self-contained, there is no obvious process that will derive it, and quite a few extrapolations are necessary to get there.

The story is told in what we would call sequential time, the events as they are experienced by the time traveler himself as he moves through history, returns to repeat it, and then repeats it again. We, however, will begin with temporal time, constructing the sequence of events as they happen in the final history as we see it, the only history if you accept Fixed Time Theory, the stable history if you prefer Replacement Theory.

The Final History

Hector and wife Clara are moving into their new home, unaware that there is a compound barely within sight on the other side of the trees where someone is working on a time machine. Yet as Hector—henceforth to be known as Hector 1—and Clara are relaxing in their yard, Hector 3 emerges from the tank that is the time machine. He has already been Hector 1 and knows what Hector 1 is going to see and do; he has also been Hector 2, who is about to emerge in under a minute. He tries to get the surprised and confused technician informed as to what is happening, but does not want Hector 2 to realize that he is there. He hides while Hector 2 also arrives the same way.

The technician is appropriately surprised. He only just activated the machine for the first time, and he has no idea who Hector is. The machine has the limitation that it can only transfer people between points during which it is active, that is, you cannot travel to a time before the machine existed, and so it should not be that surprising that someone has emerged from the machine shortly after it was activated; but the researcher (whose name is never given) is reasonably surprised that he does not know the human traveler. However, Hector 3 directs him to keep Hector 2 from seeing him, so he can fix what went wrong the last time.

The researcher partly explains and partly questions Hector 2 to get a better understanding of what has happened, and determines that Hector 2 traveled back about ninety minutes from shortly after dark to late afternoon. They observe Hector 1 and his wife in Hector's yard in the distance through the trees. Hector 3, meanwhile, attempts to leave the property on a borrowed golf cart, but the gate requires a remote signal to open, so he doubles back to meet the researcher after the latter leaves Hector 2 in the house. Hector 2 steals the researcher's white car and leaves the property, using a remote on the key ring. Hector 3 pursues in the compound's pickup truck.

Hector 2 sees an unnamed young woman pass on a bicycle, and recognizes her from when Hector 1 found her in the woods, still to come. This causes him to stop abruptly, skewing the white car awkwardly in the road. Hector 3 anticipates this, stops the truck so as to hide himself from the woman, and then continues, intentionally ramming the white car off the road into a ditch, but losing control of his own vehicle, colliding with a trash can, and going off the road in the opposite direction, out of sight. Hector 2 has a head injury, and unwraps a bandage from his arm to wrap around his head. The blood of his head wound mixes with the white liquid that fills the time travel tank, turning the bandage pink, and he wraps it around his head several times until all is obscured but his eyes and mouth.

The woman apparently hears the crash and returns to offer help to Hector 2. She produces a pair of scissors with which she trims his bandage, and tucks it so it will stay securely. Her cell gets no signal so she says she is going for help, but he persuades her to stay, and then to come with him. He palms her scissors, and then uses them to threaten her, getting her to remove her shirt and replace it so that Hector 1 will see this through his binoculars. This rouses Hector 1's curiosity, but he patiently waits as Clara leaves for town to get food for dinner. Hector 2 then gets the woman to remove her pants, but she does not believe his assertions that he will not harm her, so she hits him with the pants and runs. He pursues her, and accidently knocks her unconscious when he tackles her. He carries her back to the spot where Hector 1 will find her, undresses her, and hides.

Hector 1 saunters out to seek the mysterious girl he saw through the binoculars, seeing her bicycle and the trashed trashcan along the way. He finds her unconscious and approaches cautiously. Hector 2 stalks him and stabs him in the arm with the scissors. Hector 1 flees, and Hector 2 gives chase, herding him to the time machine.

Meanwhile, Hector 3 awakens and escapes the truck. He finds one of the compound walkie-talkies and tells the researcher that he's made a bigger mess of things, and that the researcher should not allow Hector 2 to make the trip he just made. Then he heads into the woods.

When Hector 3 awakens and escapes the truck, this puts all three Hectors in the woods. The woman revives at this point, and in her effort to escape Hector 2 runs into Hector 3, whom she does not recognize because he is not bandaged, she never saw Hector 2 without the bandages, and she was unconscious when Hector 1 encountered her. She screams, whether simply from being startled or because having now been through the same accident twice he is visibly badly injured. This diverts Hector 2 from following Hector 1. Hector 3 persuades the girl that she will be safe with him, and then collapses from exhaustion sitting in the woods

While Hector 2 is searching the woods for the girl, Hector 1 breaks into the compound looking for help, breaking a window into the main house. Finding medical supplies, he bandages his bleeding arm with the wrap which Hector 2 removed and put around his head, and continues exploring the building in search of help.

As the day is fading, the girl persuades Hector 3 to follow her to a nearby house, which happens to be his house. She leaves him in the kitchen and heads upstairs. Clara enters, finding the injured Hector 3, and they hear Hector 2 enter. Hector 3 takes Clara outside, closes her in the storage shed, and uses a ladder to enter through a window off a low roof. He waits at the top of the stairs, then smashes Hector 2 with a table. Persuading the girl that she must hide from him, he cuts her hair and dresses her in Clara's coat, then has her run to the attic to lure Hector 2 that direction. Hector 3 returns to the shed and takes Clara out to the back yard where they sit in lawn chairs.

Seeing the ladder, Hector 2 climbs onto the low roof, but before he descends the ladder he realizes that someone is above him; he believes correctly that it is the girl, grabs her foot, and pulls her down. She falls to her death; but with the coat and the haircut, looking down from the roof in the dark he mistakes her for Clara.

Hector 1 picks up a walkie-talkie in a lab, and when he uses it the researcher responds. Hector 2, still carrying that same walkie-talkie, hears this, and remembers that Hector 1 leaves the radio on the table when he attempts to barricade the door against the man in the pink bandage (himself), so he uses that brief time to coach the researcher to lure Hector 1 into the tank and send him to the past. Hector 2 grabs his car and drives to the compound, breaking through the gate. He appears in the window of the room with the time machine in time to frighten Hector 1 into the tank, and the researcher closes it and sends him back about one and a half hours to become Hector 2.

Hector 2 then argues with the researcher, insisting that he must be sent back to fix what went wrong. He deduces and then gets the researcher to admit that he has already arrived in the past but told the researcher not to send him, and he gets violent in his effort to persuade him. He removes the bandage from his face, gets into the tank, and travels back to half a minute prior to his other arrival to become Hector 3.

The police have two wrecked cars, both taken from the compound, and a girl who fell from a roof of a house where she did not belong. At the time the girl fell, Hector was in the company of his wife in the back yard. Their car was stolen and left at the compound up the road; it was damaged breaking through the gate. If they use the best tools at their disposal, they might determine that the truck hit the white car, but Hector was driving both vehicles. The accident occurred while he was home with his wife.

Are you confused yet?

An Original Timeline

It should be fairly evident from our harmonization of the events that most of the critical events which happen to Hector 1 are due to the affirmative actions of Hector 2, who is specifically attempting to recreate events as he experienced them. Further, several critical events in the timeline of Hector 2 occur entirely because of the intervention of Hector 3. The question is whether we can make any of these events happen by beginning with some other cause. Let us begin by considering what would happen to Hector 1 if there were no Hector 2 and no Hector 3.

Hector 1 has won a bet with Clara that she cannot get the table she just constructed through the door; she is going to town to buy food, leaving him home. In the film he is suddenly the more eager to stay home, because with his binoculars he just saw a girl in the woods remove her shirt. However, the reason the girl is there, and the reason she removes her shirt, is because Hector 2 has brought her there and insisted at scissor-point that she do so. Absent Hector 2, the girl is not there, and Hector 1 sits in his chair looking at whatever he sees in the woods.

Even were the girl for some inexplicable reason to take that path and pull her shirt up at the right time and place for Hector 1 to see her, without Hector 2 there is no obvious way for her to wind up unconscious and naked on the ground for Hector 1 to find.

Against all probability, if she happens to have taken off all her clothes and swooned so that he finds her unconscious, that's not enough. He has to flee to the compound on the other side of the woods. He does that because he is stabbed and pursued by Hector 2. Without Hector 2 to chase him, he would probably stay near the girl to find out if she was all right. So there is no reason for him to go to the compound, certainly no reason for him to break the window and enter, no reason for him to bandage an uninjured arm or to take the walkie-talkie to communicate with the researcher. If he is not trying to escape from Hector 2, he does not head for the silo; if Hector 2's bandaged head does not appear in the window, he does not hide in the tank, and thus does not make the trip to the past.

The complications are only getting started. If Hector 1 does not bandage his arm, Hector 2 does not have the bandage with which to wrap his head after the car accident. The researcher encourages Hector 1 to hide in the tank based on a lot of lies about the pursuit of Hector 2 because Hector 2 has told him to make sure Hector 1 gets in the tank, and because having seen Hector 2 emerge from the tank the researcher knows that something seriously bad will happen if Hector 1 does not get into it.

Thus everything Hector 1 does is dependent on Hector 2 manipulating him either directly or through manipulating the girl and the researcher. Without Hector 2, Hector 1 spends a quiet afternoon in his back yard and has dinner with his wife when she returns with the food. If he does this, Hector 2 never exists.

But actually, it's worse than that.

The First Time Traveler

Everything Hector 1 does in the movie is the result of manipulation by Hector 2, who draws the girl into the woods, stages the performance for Hector 1, chases him to the compound, and conspires with the researcher to get him into the tank which is the time machine. It is worse than that, though, because from the moment he has the car accident, much of what Hector 2 does is done solely to recreate the events he remembers having experienced as Hector 1. He brings the girl into the woods, positions her where she will be seen, and has her remove and replace her shirt. After she is unconscious, he strips her clothes from her and positions her where he remembers finding her. He communicates with the researcher when he knows that Hector 1 is away from the radio, and even appears in the window in a timed effort to frighten his self into the proposed protected hiding place.

The necessary actions of the girl and the researcher are improbable absent the manipulation by Hector 2. What compounds it is that the manipulation by Hector 2 is only plausible as an effort to recreate the events as he perceived them as Hector 1. He has no reason to direct the girl to remove her clothing but that he saw her do so; no reason to stab Hector 1 but that his own arm is wounded; no reason to appear in the window but that he saw himself in the window; no reason to pressure the researcher to cooperate but that he remembers what the researcher did before.

Thus not only will Hector 1 not do what he does without Hector 2 to manipulate him into doing it, Hector 2 will not manipulate him to do it unless he remembers having been manipulated in exactly those ways. The relationship between the actions of Hector 2 and Hector 1 are so intricately interdependent that there seems no plausible way for either of them to do what they do absent the acts of the other.

And yet it gets still worse: the actions of Hector 2 are in turn controlled by interference from Hector 3.

The Second Time Traveler

Hector 2 initially is sitting in the compound house, but he's uncomfortable. It bothers him, inexplicably, that Hector 1 is sitting in his yard talking to his wife. That would be a bit like me being jealous of the fact that I was talking to my wife an hour ago, and I'm not now, so that earlier me had an opportunity I do not now have; but I suppose the fact that I can't see my earlier self talking to my wife makes it different. In any case, he steals a car to escape the compound.

He stops when he sees the woman pass on her bicycle, because he recognizes her as the naked woman he as Hector 1 found in the woods. He is probably wondering, as we did, why she would be naked unconscious in the woods. He stops the car in an awkward position, and is hit from behind and knocked off the road. It is at this point that he removes the bandage from his arm and wraps it around his head.

But this was not an accident. Hector 3 knew where he was stopped, and intentionally rammed him. Had he not done so, the girl would not have returned to the scene of the accident, and Hector 2 would not be wearing the bandage. Absent the bandage, the girl is not going to produce the scissors, which Hector 2 needs to stab Hector 1 and to threaten the girl into cooperation; further, if he is not wearing the bandage, she is going to recognize Hector 3 as Hector 2, and is not going to trust him.

Hector 3 is not finished manipulating Hector 2. Fearing that Hector 2 is going to kill Clara, he hides Clara in the shed and then uses the girl, disguising her to look like Clara, to lure Hector 2 upstairs and onto the roof from which Hector 2 drops the girl and thinks it was Clara. This leads Hector 2 to decide, after chasing Hector 1 into the time travel tank, that he must travel back to prevent himself from killing Clara.

Without Hector 3, the girl almost certainly will not run to Hector's attic at the right time, and thus will not be on the roof for Hector 2 to grab, nor fall to her death. We might suppose that Hector 2 instead kills Clara, but the woman he killed was wearing Clara's coat which Hector 3 found upstairs when Clara had not gone upstairs, so Clara was not wearing that coat. Hector 2 might not have killed anyone, had Hector 3 not intervened.

Yet the only reason Hector 2 insisted that the researcher send him back was because he believed he had killed Clara, and thus he must have killed someone who looked like Clara. Otherwise Hector 2 will have no reason to become Hector 3.

The actions of Hector 1 are due to manipulation by Hector 2, and those manipulations are based on Hector 2's memory of the movements of Hector 1. Further, other actions of Hector 2 are the result of manipulation by Hector 3, whose existence is dependent on his own interference.

Hector 3's actions are a bit more difficult.

Upon arriving in the past, Hector 3 gets the researcher to work with him, as he hides and attempts to exit the property first. Failing this, he remembers that Hector 2 stops when the girl passes on the bicycle, and he intentionally rams the white car with the pickup truck. This is a peculiar move. We know that because of this non-accident Hector 2 dons the head bandage and meets the girl, and that if he is not rammed off the road he will not do what must be done to lure Hector 1 to the time machine. Yet it is not clear how Hector 3 thinks he is going to correct matters by causing them to happen as they did. It makes more sense to suppose that having fastened his safety belt Hector 3 thinks that crashing into Hector 2 will prevent the latter from interfering with Hector 1; yet since he knows that Hector 2 is going to be stopped there, he must also know that getting hit from behind is what starts Hector 2 on the course that creates all the trouble. It would make more sense for Hector 3 to position himself behind Hector 2 so as to prevent an accident. On the other hand, he does not have time to think through everything, so he might not realize that he is falling into the history he remembers.

Once he revives, he is without a plan. Thus he wanders into the woods trying to think of one. His meeting with the girl is fortuitous, but he still does not have a plan even when she brings him to his own house. It is not until he finds his wife that he decides what to do. That decision is based on the fact that he knows that Hector 2 will accidently kill someone he believes to be his wife, but that the girl and his wife are sufficiently similar in appearance that with very little effort he can save his wife's life by replacing her with the girl.

This last part is quite interesting. He does not know that his wife died; he believes she died, but he did not get a close look at her. If indeed it was his wife who died, he is attempting to change history in essence by cheating the past into believing that the same thing happened because someone else who looks the part filled the role. Surprisingly, this actually could work in a replacement theory story: as long as the person who does not die does not thereafter interfere with necessary events and the person who dies instead is not necessary to those events, history can stabilize in the altered form. It would in fact not be necessary for anyone to die, except that in order for Hector 3 to exist Hector 2 must believe that Clara died.

It thus seems fairly evident that none of the events of this film are at all likely to have happened as we see them, because on some level each iteration of Hector bases his actions on the actions of the other two. However, before we surrender the film as impossible, it is worth considering whether any of the events are at all plausible.

The Girl in the Woods

We observed that our nameless girl removes and replaces her shirt at the insistence, at scissor point, of Hector 2, so that Hector 1 can see her, and then is found naked and unconscious in the woods by Hector 1 because of Hector 2's actions. We suppose that this only happens because he makes it so. Yet we might have an alternative for this.

We are never told where the girl is going or why she is bicycling down that road. At one point she tells Hector 2 that she is supposed to be somewhere, but she does not say where and by that point she is attempting to escape, so it may be a lie. There is a well-worn path to a sunny clearing in the woods, and it is visible only from a house which from its appearance is not quite finished and thus new, never previously occupied. Perhaps our girl is aware of the clearing and when she has time on sunny afternoons goes there to do a bit of secluded nude sunbathing. On this particular afternoon, unaware that she was spotted through Hector 1's binoculars, she makes herself comfortable on the leaves and falls asleep.

There is the problem that in later timelines she rides past the path without stopping. However, we only know that she does so when as she approaches it a car, driven by Hector 2, is passing in the opposite direction. It is not at all unlikely that she would not wish to be seen by a stranger headed into the woods alone. He might be the property owner, and might follow her to tell her to get off his land. He might be the sort of person who would follow a lone girl into a secluded wood for undesirable purposes.

She wants her privacy; she won't let anyone know that she is headed there.

We noted it is a well-worn path; someone uses it fairly regularly. It may be a spot she and her friends use specifically for this purpose, or sometimes a secret rendezvous location for young lovers. The path does not appear to serve any purpose other than to bring people to that clearing. It is too big to be a game trail, too flat to be a drainage bed.

So we could have an original history in which the girl strips naked in the woods and is seen through Hector 1's binoculars so that he is lured into the woods to find the girl. Perhaps she screams—she is asleep, not unconscious—and he runs and becomes disoriented in the unfamiliar woods.

This also means that when Hector 2 gets in an accident at that location, he interrupts her sunbathing plans and undoes the cause of her presence there. Thus it becomes necessary for him to force her to do what he saw her do so that his own existence, not to mention time itself, is not threatened.

This does get Hector 1 out of the yard and into the woods, and possibly to the border of the compound. The problem is, it does not get Hector to the time machine. He will flee when the girl screams, but when he runs into a barbed-wire-topped fence he won't scale it, but instead will follow it around the property, find the drive, and trace his way back to the road and home again. We need some cause for him to scale the fence and seek refuge inside the compound. The girl does not give us this.

What we need is something or someone chasing Hector. The problem here, though, is whence they came and whither they went. That is, if there was someone chasing Hector 1 in the original history, why do we not see him when Hector 2 replaces him? Whence did he come in the original history that he does not appear in the replay, or where does he go in the replay?

There is a potential solution. There must have been an original attacker. He was aware of the girl's route, and knew that she would bicycle through that blind corner by that path to the clearing, so he lay in wait for her there. Perhaps knocking her from her bicycle, he then threatens her, forces her up the path into the woods, forces her to undress, and renders her unconscious.

Thus Hector 1 sees the girl remove her shirt and becomes interested, not seeing the assailant, and he wanders up that direction. Whether the attacker took his time or was still there for some other reason, he is interrupted by Hector 1 blundering into the scene. He hears the man coming, hides, and then using whatever weapon he had stabs Hector 1 in the arm.

He probably did not mean to stab him in the arm. He probably meant to catch a kidney or land some other fatal blow. He does not want to be revealed, and he intends to leave Hector's body hidden in the woods probably along with that of the girl. But he missed as Hector took the blow painfully but not fatally in the arm and ran. The attacker now has to chase Hector, because he cannot let the man escape. He has been seen. Yet if the assailant knows the girl's route and the location of the hidden glade, he probably also knows that the compound Hector enters is closed for the weekend, and is rather difficult to enter or escape. He has to worry about the girl, too. He returns to her, disposes of her, and then returns to see if he can finish Hector. He might have a car hidden off the road (he had to get there somehow and have plans to escape), and he might have tools in the car.

We know that the researcher lied about having security monitors throughout the compound. It must therefore be the case that Hector sees his attacker approaching, probably along the lit path the researcher provides to the silo. The researcher might also be frightened, but he probably is thinking that he can hide Hector in the tank and hide himself somewhere else (if he also hides in the vat, who will open the vat?). He may have meant to drain the vat, but then on impulse sent Hector to the past, to see whether it would work or to remove him from danger.

This is an important point. The researcher does not know whether the time machine works. He activated it less than two hours ago, and has not run a single test. He ought to test it with inorganic materials first; but he shouldn't be here right now anyway, and if this other man who shouldn't be here is willing to climb into the tank to save his life, and the researcher risks the man's life to test the machine, who will ever know if it fails? The researcher sends Hector 1 back ninety minutes.

What happens to the researcher? Does the attacker break in and kill him? Nothing happens to him, because at the moment Hector 1 leaves from this original history, the original history ends. All pieces are restored to their positions at the time of Hector 2's arrival, and our assailant will now react to the changes Hector 2 makes in history. The problem is why that pursuer is not involved in the history we see. The answer is actually already present in the problem.

We know that Hector 2 was not comfortable sitting in a building awaiting the arrival of Hector 1. Had he done so, presumably the attacker would have forced the girl into the woods, Hector 1 would have seen her and meandered out for a closer look, and the attacker would have stabbed him and chased him to the compound and ultimately to the lab. At that point, Hector 1 would have made the same trip to the past, becoming and thus replacing Hector 2, and early that evening immediately following his departure the attacker would have burst into the lab, questioned and likely killed the researcher, and continued his search for Hector, who is now Hector 2 hiding at the house below. The girl is already dead, her body hidden in the woods.

However, for some reason Hector 2 is jealous of Hector 1. He does not like the fact that his temporal doppelganger is living his life, despite the fact that it is the life he already lived. He steals a car, leaves the compound, and as fate would have it happens to be driving through that blind corner at the exact moment that the girl is passing on the bicycle in the other direction. Our attacker has missed his chance; the girl is gone.

Now, though, we have a new problem: the girl is indeed gone. Neither the theory that a rapist attacked her there and forced her into the woods nor the theory that she was headed up that lonely path to do some nude sunbathing will get her to the clearing. Further, we have no car crash. Although Hector 2 has stopped the car abruptly in the road, there are no other cars coming—if there were, they would at least have passed the crash in the next timeline. Without the sound of the collision, the girl does not return to the scene.

It must be that Hector 2 stops the car, thinks for a moment (as he does in the timeline we see), and then, not being hit by Hector 3, turns around to pursue the girl, and somehow persuades her to come with him. He does not have the scissors, but instead must find a sharp implement in the car—perhaps a pocket knife, a boxcutter, or even a screwdriver. She sees his face, because he is not bandaged and there's no reason for him at this point to be bandaged; but since she is not going to meet Hector 3 in this timeline and she will be unconscious by the time Hector 1 encounters her, that's of no consequence. However, Hector 2 will probably use something to hide his face when he is awaiting Hector 1, because he will think it more likely that Hector 1 will flee from a masked man than from a doppelganger. The bandage is the most obvious choice, and has the advantages that it will draw attention to the masked face and away from the familiar clothing, and that Hector 1 will not recognize it as his own. It will not, however, turn pink, as the amount of blood in it from the arm is minimal and there is no head wound from the accident.

Hector 2 thus lures Hector 1, fights with the girl accidentally rendering her unconscious, and chases Hector 1 toward the compound. He then doubles back to try to explain things to the girl and make sure she is all right, but by then she has left in search of help. That leads to our next problem: who falls from the roof?

The Woman on the Roof

We considered the circumstances that led Hector 2 to bring the girl into the woods when Hector 3 was not involved to cause the crash. We reached the point at which Hector 2 has chased Hector 1 toward the compound; Hector 1 will scale the fence, break into the main building, and look for help, because he believes he is being pursued by someone intent on killing him. Meanwhile, Hector 2 doubles back to attend to the girl, but she has already awakened and fled the scene.

Hector 2 has good reason to want to explain things to the girl. After all, in this history she saw his face, and she could very well be headed for the police to report that she was taken prisoner, forced at knife point to remove her shirt, and then attacked, after which she awoke naked in the woods. She was not raped, but they might assume it was a sexual assault of some type based on her testimony. He needs to explain something to her so the police won't come looking for a man fitting his description. Even if they never arrest him, it's going to seem strange to his wife.

As for the girl, she is looking for help, and although the timing will be different she will find his house, which he left open when he wandered out of the yard, probably before his wife returns from town. She will enter in search of help. Since the house is open, she will assume someone is home and look around a bit.

Hector 2 is probably on much the same schedule as we saw, so he will arrive moments after his wife gets home. Without Hector 3 to direct her elsewhere, Clara is likely to believe that the masked man approaching is a threat—particularly if she has already connected with the girl, who will tell her that there is a madman in the woods pursuing her. Seeing the approaching masked man, Clara will tell the girl to hide, and she also will hide.

The next piece is tricky. Hector 2 believes that the woman he pulls from the attic who falls to her death is his wife. In the final history, it is actually the girl, dressed in the wife's coat and having her hair cut short by Hector 3. In this history, though, Hector 3 is not there to cut the girl's hair, put her in the coat, or send her to the attic; nor did he stash Clara in the shed. Yet if Hector 2 does not believe that Clara fell to her death, he will not insist on making another trip to correct things, and Hector 3 will never exist. Therefore there are only two possibilities:

1. The girl hid in the attic and fell to her death, but despite the fact that it looks like the girl and is dressed like the girl and Hector 2 expects it to be the girl, he mistakes the dead form for his wife. This is unlikely.

2. Clara hides in the attic and falls to her death.

Given these choices, it seems the latter must be the case. Hector 2 thus realizes that he has ruined his own life, and now determines to change it. The movie poster is now right: he does change history.

But first, he has to finish chasing Hector 1 into the tank. Hearing Hector 1 and the researcher on the walkie-talkie, he remembers setting it down and walking away, and so contacts the researcher and says he is coming. He grabs the keys from beside the door and takes his car to the compound, popping into view in the window in his bandage mask to frighten Hector 1 and finish the job.

Then, because he believes he has killed Clara—which he has—he demands to be sent back again to fix things. The researcher did not receive a message from Hector 3. He is still reticent; he does not like the fact that something might have gone wrong, and his time traveler seems to be something of a loose cannon. But he agrees under pressure, and so Hector 3 is born.

The Third Hector

Hector 3 starts with one objective: prevent Hector 2 from luring Hector 1 to the time machine. It would be fatal if he were to succeed, because he would undo his own existence pushing everything back to the lost original history in which Hector 1 wound up in the time machine—an infinity loop. However, Hector 3 has not thought that far ahead.

At first he intends to beat Hector 2 to the girl, but the gate is locked against him and by the time he can get a vehicle with a gate key Hector 2 is already ahead of him on the road. He next thinks that he can manage this if he crashes the pickup into the white car, injuring Hector 2 and so keeping him from pursuing the girl, perhaps again intervening such that he meets the girl instead, and sends her to safety. This also fails, because the girl finds Hector 2 and not Hector 3, and so Hector 2 starts manipulating her into his plan while Hector 3 hangs unconscious nearby in the inverted pickup truck.

By the time Hector 3 awakens, Hector 1 is already headed for the compound with Hector 2 in pursuit. Upset that his plan to prevent Hector 2 from catching the girl failed, he radios the researcher and tells him not to allow Hector 2 to become Hector 3. It is unclear why Hector 2, who is carrying the walkie and walking in the quiet woods listening for some sound from Hector 1, does not hear this, but apparently he does not.

Fortuitously, when Hector 3 goes into the woods the girl happens upon him. Her scream diverts Hector 2 from his pursuit of Hector 1, but Hector 3 is right that they are hidden where Hector 2 won't look. In this timeline, Hector 2 has been bandaged from the moment she saw him, so she does not recognize Hector 3 and is easily persuaded by the badly injured man that they might both be fleeing from the same attacker. She eventually persuades him to come with her to the nearby house—his house—for help, and although he attempts to dissuade her, she is determined and he is injured, having twice been in the same automobile accident and sustained head injuries both times.

When the girl leaves him in the kitchen and Clara appears returning from the store, he gets his idea. He is going to attempt to lure Hector 2 away from Clara, using the girl and the ladder as decoys. He probably hopes that Hector 2 will hear the girl, then see the ladder and think that the girl got from the attic to the roof and down the ladder. However, he is also hedging his bets: he disguises the girl to look as much like Clara as he can manage in a few minutes, so that if the girl falls Hector 2 will think it was Clara. He then moves Clara to a safe location, settling next to her to establish his own alibi for the events to come.

The plan works quite well. The girl falls to her death, and Hector 2 mistakes her for Clara. At that moment the walkie-talkie reminds him that Hector 1 needs to be chased into the tank, and he grabs the car keys and heads for the compound. After that he, in his turn, becomes Hector 3.

There is a confusing aspect to the actions of Hector 3.

It makes perfect sense for Hector 3 to think that if he crashes the pickup into the white car, he might prevent Hector 2 from pursuing the girl in the car as he did when there was no Hector 3. However, once the Hector 2 who was knocked off the road by an unseen vehicle becomes

Hector 3 and finds himself stopped on the road around the bend from where he knows the white car is stopped, he must know that the white car was hit as he intends to hit it, and that the consequence was that the girl returns to the scene of the accident to help and is taken prisoner by Hector 2. Why, then, does he go ahead with the collision that he already knows will not have the impact he desires?

It must be noted that he does not know that the white car was hit by the pickup truck. He only knows that it was hit by something. Perhaps he hopes that if he does this right, instead of the girl coming back to the white car she'll come back to the pickup truck, and he can send her on her way without incident.

It also must be noted that this was not his original plan. When he emerged from the tank, it was his intention to get ahead of Hector 2, and thus probably to divert the girl before Hector 2 could catch her. That failed; he is seeking another plan. When it occurs to him that Hector 2 is parked around the bend ahead of him, it might not occur to him that he was the original cause of the crash and that it did not work as he had hoped. Thus without time to think through the likely consequences, he secures his safety belt and rams the white car, only later realizing that he has caused the history he hoped to prevent.

When they come to the house, he is ambivalent. On the one hand, he knows that someone died at the house when he was Hector 2, and he thinks it was Clara, and he thinks that he, believing she was the girl because he'd traced the girl to the house, accidentally killed her. Thus it might be that if the girl never goes to the house, Hector 2 will not go there either, and Clara will be safe. On the other hand, it may be that Hector 2 will mistakenly believe that the girl went to the house and will frighten and chase Clara. Thus

Hector 3 forms the notion of replacing Clara with the girl—a plan which coalesces when Clara finds him in the kitchen and he realizes that he can save her by sacrificing the girl.

At this point we have a self-sustaining stable history containing two anomalies both of which have resolved into N-jumps. A great deal of extrapolation has been needed to work out what must have happened to get us to this point, but the only serious coincidence is that there must have been an original attacker who dragged the girl into the woods, and that this attacker was frightened away when Hector happened to be at the point of the intended ambush at the same moment as the girl.

Congratulations are extended for a well-executed story.

Analyzing Predestination

When this book was in what might be called the outlining stages, temporal anomalies fans were asked what movies should be included. Several were suggested, but the popular one was *Predestination*, based on a Robert Heinlein story (*All You Zombies*) involving some very complicated predestination paradoxes.

This was a fascinating and challenging movie presented with Ethan Hawke in the lead role, with a supporting role by Noah Taylor who also had a significant supporting role in the time travel film *Edge of Tomorrow* (as Doctor Carter), making this his second time travel movie. It is built on a predestination paradox, or perhaps several such paradoxes, and may be the most convoluted tale of narcissism ever devised. It is well worth seeing; I am reliably informed that it is worth seeing even if you have read the original. Unfortunately, it is not remotely possible to discuss the temporal elements within it without major spoilers, and so if you have not seen it you perhaps may wish to view a copy before continuing with this chapter.

Hopefully you have now seen the movie.

Temporal Order

The first challenge is that the story is not presented in either chronological order (that is, following the calendar) nor sequential order (following the events of the time traveler's experience), but in an order organized around events while maintaining the surprises until later. Thus it may help if we begin by putting all the major events in the order in which they occurred, complete with such dates as are provided or can reasonably be extrapolated. However, it is still convoluted and extensive.

There is also a problem of names. The main character telling most of the story never actually gives us his, although we eventually deduce it; in the subtitles he is identified as "Barkeep", which works as a name which distinguishes him from other characters. The second critical character gives his name as "John", so although he is identified in the subtitles as "The Unmarried Mother" (his *nom de plume*) we will use "John". The girl, from baby until motherhood, is always "Jane", although sometimes it will be necessary to distinguish the "Baby", also named Jane, from its mother, "Jane". There is also an important figure known as the "Fizzle Bomber" whom we will call "Bomber" even though eventually we do learn his name.

On September 13th 1945 someone registered at a hotel as "Gregory Johnson" but who is actually Barkeep delivers a two-week old baby, Jane, to the Cleveland City Orphanage. Baby Jane grows up there wishing she had a normal family, but is introduced to Mr. Robertson, loosely connected with Space Corp, and begins training for that program. She is disqualified after a fight when it is discovered—but not revealed to her—that she is a hermaphrodite. Then, on April 3rd, 1963, she meets John, and quickly falls in love. This whirlwind romance lasts a little less than three months, as on June 24th, 1963, Barkeep comes and takes John away to enter the Temporal Bureau.

Robertson manages to get Jane back into Space Corp training, but she then discovers that she is pregnant and is dropped from the program. The delivery goes badly at the charity ward, turning into an emergency Caesarean Section and total hysterectomy—but this time when they discover that she is a hermaphrodite with hidden immature male genitalia, they work on reconstructive surgery and hormone therapy to transform her into a man.

Meanwhile, on March 2nd 1964, Barkeep kidnaps Jane's two-week-old baby Jane from the hospital nursery, to take her to the orphanage (in 1945).

Oddly, Jane takes the name "John", which is the more odd because she hates John for ruining her life, but perhaps reasonable as "Jane" is a feminine variant of "John". As the therapy has its effect she looks more and more like John, but does not realize that she is becoming him.

We have covered almost twenty years of Jane's life to the point that she has had a baby and started to become John, who will be the father of her baby, who is herself. That is the first of the several surprises in the movie; there are more in the later years, after a four-year break about which we are told little.

Sometime in 1968 the Fizzle Bomber destroys the Hardshaw Weapons Factory; this apparently prevents a terrorist occupation of that building that killed at least three hundred forty-six people. It is the earliest attack by Bomber reported to us.

At 8·45 on March 2nd, 1970, Barkeep arrives in the area of one of Bomber's attempted bombings, and goes to the location. Bomber is setting an eighteen-minute timer, and gets in a gun battle with Barkeep, which takes them out of the area. Bomber leaves Barkeep stunned, but by this point the timer is down to a couple of minutes and John has arrived to prevent this bombing. He is about to remove the bomb when he hears Bomber returning; they get into a gun battle as well, but Bomber flees in time for John to remove the bomb and put it into a containment unit—but not in time for him to seal the containment, with the result that he is badly burned by the explosion. Barkeep walks over and puts John's USFF Coordinates Transformer Field Kit (disguised as a violin case) within his reach, and John retreats to the future. Barkeep finds a piece of the bomb timer, and leaves for 1964 to kidnap the baby.

On November 6th, 1970, John walks into the bar where Barkeep is working, and tells his story, and Barkeep offers to give him the opportunity to kill the guy who ruined Jane's life if John will then consider working for his boss, who is the same Robertson who tried to recruit Jane earlier. John agrees, and is removed from that timeline. A short while later, Barkeep returns to the bar and quits his job, then again leaves for his retirement in or about 1975 by way of 1985.

Sometime in 1974 there is a major Chicago chemical spill which kills 324 people; Bomber sets a bomb which prevents the driver from getting to work that day, preventing this disaster.

Sometime very early in 1975, possibly very late in 1974, Barkeep arrives in New York intending to retire. His USFF Coordinates Transformer Field Kit shifts to "DECOMMISION" but then a moment later gives the message "FAIL ERROR FAIL". He opens the envelope he got moments (or years) before from Robertson, and finds information on the timer he found which enables him to track Bomber to a laundromat at one in the morning on a particular unidentified day in March.

Before that day he meets Alice at an antiques shop; this apparently develops into some kind of relationship, because Bomber ridicules it when they meet.

At that meeting, Barkeep realizes that he is—or will be—Bomber. Bomber wants to restore their past loving relationship and be together; Barkeep is horrified at the explosion that has not yet happened. Barkeep kills Bomber.

A few days later, eleven thousand people are killed when the Fizzle Bomber levels ten blocks of New York City—or that is the history Barkeep was trying to prevent and should have prevented by killing Bomber.

The rest of the chronology is something of a denouement. In 1981 time travel is invented, with a range of plus or minus fifty-three years from the moment of its invention. 1985 is in some way the official "headquarters" of the Temporal Bureau, and on August 12 of that year, at 11:01 in the evening, Barkeep arrives with John, who is about to become a temporal agent. On April 3rd, 1991 Bomber destroys something in Hamburg, Germany, preventing a history in which one thousand eight hundred sixty-one lives were lost. On February 21st, 1992, a burnt John arrives from 1970, is treated, undergoes plastic reconstructive surgery, and becomes Barkeep.

So if you were keeping score, Jane, John, Jane and John's baby Jane, the Barkeep, and the Bomber are all the same person interacting with himself over a span of about half a century. Next we will attempt to put it together in the order in which he experienced all this.

Birthing Baby Jane

Jane is born on or about February 17th, 1964, given that she was two weeks old when she was kidnapped on March 2nd, and 1964 is a leap year. Two weeks later she is kidnapped and taken to September 13th, 1945, where she is left at the orphanage. She is largely undisturbed by time travel until June 24th, 1963, when she is almost eighteen years old and meets John. She gets pregnant and he vanishes, and she is forced to drop out of her training program and have the baby in the charity ward. The baby is kidnapped, and she is badly harmed by the caesarean section; she is informed of her hermaphroditism and surgically and hormonally altered into a man, becoming John; although she looks familiar to herself, she does not realize that she is the same person who becomes the father of her baby.

She, now he, becomes the moderately successful writer of *Confessions* stories under the *nom de plume* "An Unmarried Mother". On November 6th, 1970, about the time the doctors announce that his male genitalia are fully functional, he meets Barkeep at Pop's Place and tells him his story. Barkeep offers him the chance to kill the guy who did this to him, claiming that he suspects that guy might be Bomber, who has by now begun to make New York City people very nervous. John accepts the offer, and winds up back on April 3rd, 1963, where he discovers that he is the guy who ruined his life, and falls in love with him/herself, that is, Jane, staying with her for most of three months and impregnating her before Barkeep returns on June 24th, 1963 and takes him to August 12th, 1985, to become a temporal agent. At this point Barkeep tells John that he, Barkeep, is John's future self. John apparently has several successful missions which are not detailed, and then is sent to March 2nd, 1970, to prevent a bombing. He is attacked by the Bomber, but manages to remove the bomb, almost containing it in the containment unit but being badly burned by the blast when it detonates. Barkeep appears, pushes the USFF Coordinates Transformer Field Kit into John's grasp, and John leaps forward to February 21st 1992, where he is immediately hospitalized and undergoes treatment and surgery to rebuild his badly burned appearance. When he is fully recovered, he is Barkeep.

As Barkeep, he now has his final, and rather complicated, mission. He takes a job as a bartender at Pop's Place a couple weeks before November 6th, 1970, so that he will be there to meet John, to take him to April 3rd, 1963, so he will meet Jane and fall in love. He makes an unauthorized side trip to March 2nd, 1970, to try to stop Bomber before the bomb harms him, but only succeeds in helping John reach the Field Kit. Then he goes to March 2nd, 1964, to kidnap baby Jane and give the piece of the bomb timer to Robertson, and begins recording instructions for John. He takes baby Jane to the orphanage on September 13th, 1945, and calls the orphanage, and then goes to June 24, 1963, to reveal his identity to John and take him to August 12, 1985, to become a temporal agent.

At this point he officially retires, choosing to live in New York City in 1975 just before the huge explosion. Robertson gives him an envelope with information about the piece of the timer he found. Several things happen in what is a somewhat overlapping order, including that the Field Kit fails to decommission, Barkeep meets Alice at the antiques store, he studies the information in the envelope, and he identifies a time and place where he can catch Bomber.

He discovers that Bomber is an older version of himself, but is so horrified at what he apparently became he shoots himself, that is, Barkeep shoots Bomber, several times. However, Bomber was clearly considerably older than Barkeep, and still had access to a time machine. We know that he made quite a few trips, in no identifiable sequence, including one to prevent a 1974 chemical spill in Chicago, one to prevent an unexplained disaster in Hamburg in 1991, and one to destroy a weapons factory in 1968 so that terrorists could not take it.

It also seems that he caused the destruction of ten blocks of New York City a few days after he is shot; that is one of the problems that must be addressed in resolving this film.

Original History

I was tempted to reserve the problem of the baby for last, but it really is in the way of everything else. The story is built on the fundamental predestination paradox that Jane is her own child. That is, in order for Jane to be born, Jane must give birth to herself. As we have said, anything that can only happen if it happens will never happen. However, such paradoxes can be resolved in Replacement Theory by extrapolating an original cause. Following Sherlock Holmes' advice, then, once the impossible is eliminated, whatever remains, no matter how improbable, must be the truth.

In this case, we simply need an original baby. Someone had a child and abandoned it at the orphanage late in 1945. The child need not have been a hermaphrodite; we will get to that. This is our original Jane, and she excelled in many of the skills we see in the Jane we meet, and pursued a similar course in life.

It then comes to the point where the child, presumably named Jane, falls in love and gets pregnant. There is a narcissistic element in the relationship we see, as she falls in love with her older male self, but as we saw in *About Time*,[27] there may well be a sense in which people fall in love with someone when they reach a moment in their lives when they are ready to fall in love with someone, so without John, Jane very likely would have met someone else right around that time, maybe at that spot. She is actually a very pretty girl, and one who is unaware of how pretty she is and suffers from self-esteem issues so would be easily flattered by the first person who actually notices.

From this point forward, it is difficult to reconstruct with any level of certainty, but since time travel will not be invented until 1981, no one from the future can tamper with events in 1963 and 1964. We presume that she gets pregnant, because that is a necessary event to a later history which will not to this point have been altered. Maybe the father leaves her before she knows she is pregnant, or after she tells him, or when the difficult cesarean section leaves her sterile. She winds up a single mother. Perhaps she becomes a writer. In any case, Robertson, probably part of the space program, will see that someone with great potential got a lot of bad breaks, and will remember her.

It then all changes in, or shortly after, 1981, when Robertson takes charge of the new time travel program. He needs agents, and he remembers one person who might

---

[27] Analysis available on the web at http://www.mjyoung.net/time/about.html. In short, in the film Mary meets Tim and falls in love with him, but then Tim alters history erasing that meeting and within a few days she meets Rupert and falls in love with him instead, so Tim has to find a way to alter history so that she meets him and not Rupert.

have been very good at this had her life not taken such bad turns. Of course, in 1981 that person turns thirty-six years old, and a younger agent would be a better recruit—but then, the life of this particular individual is such that it would be difficult to find the right time. By the time she was eighteen she was pregnant; by the time she was twenty she was a single mother trying to raise a daughter who would undoubtedly face difficulties herself. Of course, the daughter also shows promise, but is sixteen going on seventeen in 1981, so it would mean waiting a couple years for her. The ideal would be to bring Jane into the program a few years after the birth, when she is about twenty-five, but the baby Jane is in the way. So Robertson has an idea, and this changes everything.

The interesting solution would be to remove the baby from the picture. Don't kill her—she has potential, too, as daughter of the promising Jane. It might be interesting to take the baby back in time, so that she will be just the right age to be an agent around the same time that Jane is the right age. Then Robertson might get two promising agents from the same orphanage, who knew each other and knew how to work together. So Robertson goes to 1964 and kidnaps baby Jane.

This creates an anomaly because history has been altered, but before that can resolve Robertson has left 1964 and arrived in 1945, creating another anomaly: now there are two foundlings at the Cleveland Orphanage.

This is a problem. We assume that the original Jane was delivered there, and now the replacement Jane is also delivered there. Where is the original Jane?

The answer is rather simple, really. Jane tells us that she was one of several. If baby Jane was delivered first and given the name Jane, then mother Jane would have been given a different name—there is no reason for an orphanage naming its babies to give the same name to two children, as that only confuses matters. Neither of them has any reason to think they are related, let alone mother and daughter; they look similar, but not identical, because baby Jane is the product of a relationship with our unknown father, and so carries his genes. If baby Jane excels beyond what her mother had accomplished, then the overshadowed original is likely to withdraw from the competition at an early age and become considerably more "normal", less impressive, and perhaps more adoptable. The original Jane vanishes into the population, replaced by her own temporally-displaced daughter.

Ah, but is Jane not a hermaphrodite? We noted that problem, and it is answered in the identity of the mystery father. Thus the original Jane can plausibly be replaced by her own daughter. However, things are about to become considerably more complicated.

To reach the point at which Jane is replaced by her own daughter, we had to extrapolate that she fell in love with and got pregnant by someone else before meeting John, and that Robertson kidnapped the baby in order to make Jane available for the time travel program. That gives us several problems related to this original father.

Our second Jane, daughter of our original Jane, takes over that history, undoubtedly with some changes but still excelling at everything. She becomes the space program candidate, and she meets and falls in love with the man whom she does not know is her own father. She also gets pregnant, he again leaves, and Robertson again kidnaps the baby to take to 1945. Since now the previous replacement baby was never born, this baby replaces the second baby Jane and in turn displaces the original Jane—but we hit a problem, a serious genetic issue that results in a cycling causality, a sawtooth snap.

Here is the problem. The original Jane had a mother and a father, whom we will call Mom and Pop, each contributing 50% of Jane's DNA. When she has baby Jane, that child's DNA is 50% hers, which means 25% Mom and 25% Pop, plus 50% that of the father, whom we will call Dad. That Jane has just met Dad and had a replacement baby Jane, whose DNA must logically be 12.5% Mom, 12.5% Pop, and now 75% Dad. (This is why incestuous relationships are genetically dangerous.) That child replaces the mother, in turn meets Dad, and has the next iteration of Jane, whose original DNA from Mom and Pop has dropped to 6.25% each, the remaining 87.5% coming from Dad. This shift away from the DNA of Mom and Pop in favor of Dad continues, each iteration reducing the original parental contribution by half and increasing that of Dad to provide the difference. Gradually Jane becomes her own father. Eventually, as her DNA approaches 100% Dad, she has to become male.

Perhaps, though, there is a solution to this that also resolves the other problem. What if the man with whom she falls in love is the hermaphrodite? While it is normal for hermaphrodites to mature as female first, it is not always the case, and environmental factors (such as exposure to testosterone products at an early age) can tip the balance the other direction. This simplifies the story: the original Jane was an extraordinary ordinary girl, and at some point as her original parents' DNA is replaced by that of her lover she becomes a hermaphrodite. Not maturing in the same environment she presents originally as female, and so still makes the same connection, and the child of two hermaphrodites is far more likely to be a hermaphrodite than otherwise.

Mathematically we still have the problem that no matter how many times we reduce the original parental DNA by half it is never zero; functionally, though, we reach a point at which the last gene has been replaced, and Jane and her lover are the same person, one presenting as female and the other as male.

If the conception process were entirely random, this would still be a problem. After all, if you have a hundred pennies and a hundred dimes and you randomly select fifty of each, you will not get the same set of pennies and dimes twice in a row. However, genes are arranged in chromosomes, which come in pairs, and the genes that can fall in any particular point in the code can only come from the equivalent position in the parental code. We are dealing with an extreme improbability—but once we reach the point at which Jane and her lover are the same person, we have at least the chance that they might produce a child who is a genetic match to themselves, and if they happen (against tremendous odds) to do that, history stabilizes into an N-jump as the exact same choices will lead to the birth of the exact same child, and history can continue—at least for the moment.[28]

---

[28] In the film *About Time* when Tim makes a minor change to the past, he alters the gender of his child. That is a possibility here, but our assumption is that when exactly the same person makes exactly the same choices at exactly the same time, we get exactly the same result. The genetic code of a child is not "random" in the ordinary sense, but only in the sense that we cannot readily predict it. The exact same causes will produce the exact same outcome, and even with the butterfly effect we can expect the same child to be born from the same chain of events.

We have managed to reach the point at which Jane is her own baby, and a hermaphrodite, and the baby was kidnapped and taken to 1945 to become her. At this point, the delivery is such that she becomes that man, and Robertson recruits him into the time travel service.

John, however, is pretty upset about that guy, whom we named Dad, who ruined Jane's life. There are rules, of course, and there is some way for those who enforce the rules to be aware of infractions (somehow Robertson knows that Barkeep made an unauthorized trip to try to catch the bomber in 1970), but that probably will not prevent John from trying to remove this pain from his past. He travels to April 3rd, 1963, and waits at the place where Jane met Dad. He intends to kill his former lover.

So what happens to the lover? Obviously, Jane encounters John first, and John being genetically exactly the same person as Dad and the future version of Jane, he becomes the replacement lover. In this first encounter he may know, or at least think, that he is not the original villain of this story, and instead may suppose that his attentions toward Jane will guard her from that scoundrel—and in some sense correctly so, but that now he is forced to leave. Probably Robertson realizes what is happening, and comes to remove John from the situation. Robertson will have to point out to him that if he does not leave, he will never become the person who arrived, so his presence here now is dependent on his departure.

This is of course problematic as well. Once it has happened, it would not be a matter of preventing John from meeting Jane but of undoing that encounter, and there is no way that Robertson can as much as identify the original Dad, let alone arrange that meeting with Jane, at that point. On the other hand, if he waits for the baby to be born John will undoubtedly prevent the kidnapping. The only solution at this point is to pull John out of time after Jane is pregnant but before the baby is born, probably before Jane knows she is pregnant. Robertson has to allow John to become his own father.

This is again complicated by that same genetic problem we faced with Dad: just because two people are genetically identical does not mean that they will produce a child genetically identical to them. The odds are perhaps twice as good as the odds that two siblings who are not identical twins would be genetically identical (twice as good because this child draws DNA from two identical parents, but is still drawing only half the DNA from each and so not necessarily getting the correct matching set). So again we have the extremely improbable circumstance that the baby must, against the odds, be a perfect copy of the parents. This time, too, it is made less likely by the fact that the situation changes, as we saw in *About Time* when Tim changed the identity of his daughter Posy to become his son. The first time John encounters Jane it will not be the meeting he remembered having as Jane because it is a different meeting; it is not until the next time through that John is the Jane who met John. That changes their relationship ever so slightly, and makes this unlikely in the extreme, because Jane and John must give birth to themselves in the person of the child that follows both from the new relationship they create in their first time through and in the replacement relationship they create when John potentially remembers having lived through all this as Jane.

Again, though, it is not impossible, only improbable in the extreme. Thus we could reach the point at which temporal agent John becomes his own father through his relationship with his own younger female self Jane, and we get an N-jump termination on a sawtooth snap that allows us to move forward a bit further.

Becoming Barkeep

We now come to an enigma. Somehow, John becomes Barkeep, and in doing so becomes someone that John, who already had the experience as Jane of meeting a future version of himself, does not even suspect might be a future version of himself. In the film, this is accomplished by introducing Bomber, and having one of the bombs burn John so severely that complete reconstructive surgery is needed. However, Bomber is a future version of Barkeep, starting to lose his mind from too many trips through time. If John never becomes Barkeep, Barkeep never becomes Bomber; and if Barkeep never becomes Bomber, John is not burned by the bomb and does not become Barkeep.

I have not read the book, but am reasonably reliably informed that Bomber was invented for the movie. As a footnote to this, it is suggested that without Bomber, this works as a fixed time story. I am not entirely persuaded; the very concept of a temporal police agency preventing crimes before they happen implies that history can be altered, as there would be no motivation to attempt to change events that were already right or could not be changed. That is, either everything the temporal police want to prevent never happened, or the temporal police have a one hundred percent failure rate, and either way they would not exist. However, that does not impact the movie. There is a clear statement in the movie that Bomber, at least, has changed history, and therefore fixed time does not work for the movie version.

What matters to us, though, is that there must have been some reason in the book for John to become Barkeep. It is rather unlikely that this was done to disguise his identity—at this point, Robertson has done everything that needed to be done. Robertson might assign John the task of taking baby Jane to 1945, but probably he does not need to be disguised for that—the hospital staff have never seen him, and Jane did not recognize him as herself. Evidently something happened which led to a decision that John needed to have reconstructive surgery. He becomes Barkeep, and (foolishly, perhaps, given the dangers of meeting yourself which we see in *Mr. Peabody & Sherman*, in *Timecop*, and in *Back to the Future II*[29]) is given those assignments related to his younger selves. He is now the one who recruits John, and who kidnaps the baby and takes her to 1945. It is probably his idea to suggest to John that he can use this opportunity to kill the original father, and thus he sets up his own meeting with himself (that is, John with Jane). There are a thousand tweaks to history, a thousand things that can go wrong particularly in that way that butterfly effects[30] do, but the remote possibility that everything will fall into place as we see it. Then Barkeep retires and chooses to go to 1970, for no particular reason. His Coordinates Transformer Kit fails to decommission, and he gets the idea that he can continue preventing disasters as a freelance time traveler. He becomes Bomber.

We have no notion of the sequence in which he does his bombings, nor even whether his first attempt to prevent a disaster was done with a bomb. Our best candidates for his first bombing are the 1974 chemical spill, where a bomb prevented a driver from getting to work and causing

---

[29] This will also be discussed later in the book.
[30] Butterfly effects were discussed in their own section in the chapter on Replacement Theory, above.

an accident fatal for three hundred twenty-four people (we do not know how many people were killed in that bombing), or the rather vaguely described 1991 Hamburg, Germany bombing in which almost two thousand lives were saved by a bombing about which we know nothing. If Bomber regarded these as successful, it would encourage him to keep going, and perhaps to accept greater casualty rates as successes. He becomes a target of the temporal enforcement agency, and John is assigned the mission of identifying and stopping him.

There are still some problems, though.

Barkeep is trying to catch Bomber, but our instinct is that this cannot be done, because Bomber already knows everything Barkeep is going to do—after all, he was Barkeep, and he has a good memory. But it does not work quite like that. As we saw in *Looper*,[31] the notion of the future self knowing what the past self does has an inherent flaw, which here is that the past self is reacting to the future self and so the information is changing.

Let us look at the complicated one, the thwarted bombing. In the original history there was no bombing, because Bomber did not exist; we go through quite a bit of alteration and reconstruction before we get Bomber, but eventually Bomber sets that bomb.

We see only one version of those events, although we see it from two different perspectives; there are quite a few other versions of them. As noted, in the original history there was no bomb, no explosion; in the history we see (twice), John arrives to disarm the bomb, gets in a firefight with Bomber, prevents the bomb from doing its intended

---

[31] Analysis available on the web at
http://www.mjyoung.net/time/looper.html

damage but suffers the injuries from being caught in its incendiary blast. We ask ourselves how John knew there was a bomb there, and there is really only one possible answer: there was a history of the world in which the bomb exploded, killing some unreported number of people. In that sense, John is not preventing a bombing but rather undoing one. This is perhaps the inherent problem with police time travel stories: if the crime did not happen, it cannot be prevented, and if it did happen and you prevent it, it did not happen. Barring some application of the misunderstanding of Niven's Law,[32] temporal police agencies fail.

There is, though, another problem. There is probably a twenty to thirty year age difference between John and Bomber. We know that John was two weeks old in 1945 and was removed from time in 1970, making him twenty-five years old. We know that he had an illustrious career. It is not clear when the mission to prevent the bombing occurred. On the one hand it seems as if it was the first mission on which he was sent, while on the other hand it seems that after the reconstructive surgery which followed it there was much talk about how many trips he had already made and that he was about to launch his final mission. Yet the last time we see Barkeep he was certain he would never become Bomber, and Bomber had a significant career, probably spanning quite a few years itself. The minimum gap we could imagine is about five years, and twenty is considerably more plausible. Let us settle for ten, for the sake of discussion.

Bomber sets the bomb in 1970, and it destroys whatever was targeted. No one arrives to stop it. Then at some point John is detailed to attempt to prevent this bombing,

---

[32] Discussed in its own section above in the chapter on Replacement Theory.

and he arrives. There is no reason to suppose that Bomber waited for him, and therefore John will successfully disarm the bomb. The Bomber who planted this bomb did not know John would do this, because he never was that version of John. It will take this John ten years to become that Bomber—then that Bomber will know that that John defused that bomb, and will stay to prevent that. He will get in a gun battle with John, and flee just before the bomb detonates, giving John sufficient time to remove it from the target but insufficient time to contain it. John is caught in the blast, and badly burned. He finds his time travel violin case, and returns to headquarters.

Note that Barkeep is not there. Barkeep has no reason to be there—in his memory, he removed the bomb and contained it, and there was no sign of Bomber. It will be a couple of years before the version of John who remembers being caught in the blast becomes Barkeep, and makes the trip to attempt to catch Bomber at the scene. That is also when Barkeep gives the violin case to John, changing another detail—but it must be a token gesture, because John must have reached it on his own in order for Barkeep to be there at all. That means that that Barkeep will now become the future Bomber, who will also remember that Barkeep was there.

Note that this is unlike *Next*, in which Cadillac's knowledge enabled him to dodge bullets and blows and evade pursuit:The different approaches to and effects of knowing the future are discussed below.[33] Bomber does not know what will happen, only what happened when he was that age, and as he changes his actions in response to his memory, his younger selves change what they do to counter the new actions.

---

[33] The different approaches to and effects of knowing the future are discussed below.

It is remarkable that he does not get caught—but perhaps he does, because in the next iteration of history he would know what he did wrong and be able to avoid making that mistake again. On the other hand, doing it right erases the knowledge of what happened when he did it wrong, so ultimately it has to stabilize into one version of events that are good enough, the version we see.

The exact date of the bombing in New York kept changing. This was probably because Bomber kept adjusting his schedule in response to his memory of events—even just to avoid his memory being correct. That is, if he remembers the bomb detonating on March 10th, he knows that his younger self will be expecting the bomb to detonate on March 10th, so he moves it to March 8th, or March 15th, or to another day when it will not be expected.

This automatically gives us a sawtooth snap: whatever the history was in the previous version of time, someone is intentionally altering it in this version. The only question is whether there would be method to his madness, some pattern which means that if last time it was the 10th, this time it is the 8th, and if it was the 8th it becomes the 15th, and if it was the 15th it becomes the 10th. That would create an infinity loop termination. Either way, though, as long as Bomber keeps changing the date of the bombing, the future is destroyed.

That is problematic, though. We have several times mentioned that if you erase an event, you also erase all knowledge of that event, because you made it such that it never happened. Further, as in *Back to the Future II*, if you cause an event never to have happened, you also prevent the existence of any newspaper clippings about it—Bomber's scrapbook is undone, and the wall of cuttings about the bombings maintained by Barkeep can only include such bombings as were successful. Thus Bomber can only remember when the bombing occurred in the history he knew as Barkeep, and Barkeep can only know when the bombing occurred in the only history that existed, and when Bomber changes the date the result is that the old date is forgotten, and the new history includes only people who know the new date. It is a foolish impossibility—Bomber can change the date, but Barkeep cannot know that it keeps changing, only whether it happened on the date anticipated when he traveled there to stop it, and then all the information changes and no one knows the date was changed.

The other problem, though, which confronts the viewer is that a few days before the bombing, this time, Barkeep tracks Bomber to the laundromat and kills him. This should, we think, prevent the destruction of ten blocks of New York City. Here, though, we are failing to think fourth-dimensionally. The simple solution is that Bomber has already destroyed New York City several days in the future, and traveled back to do his laundry, perhaps because this is laundry day and his decaying intellectual abilities force him to keep to the routine, perhaps because he wants to meet the younger self with whom he is still in love, perhaps merely to establish an alibi for the bombing. In any case, Barkeep fails to recognize that this has already happened, and so as he kills Bomber he thinks he has prevented a bombing which already exists in the future, already was accomplished by his older self.

It is from there a simple matter to suppose that Barkeep gradually becomes Bomber. He perceives that he can still prevent disasters, and begins doing so, and as his mind deteriorates from excessive time travel he starts to see the merit in using bombs, preventing a greater disaster by causing a lesser one. The climax of his career is, of course, the destruction of ten blocks of New York, and we have no idea what disaster he was attempting to prevent, but perhaps that one did not work as planned—and then he kills himself in a way that is not viewed as suicide, and his story ends.

Ultimately, there are some problems that are sticking points—notably the moving date of the New York bombing, also the aspect of preventing crimes or disasters before they happen based on the knowledge that they happened when they were not prevented—but overall it was a pretty decent time travel film, and what were presented as the big problems proved on consideration to be soluble, if perhaps highly improbable. It was also a fascinating movie to watch, well worth the time.

Meeting Yourself

In our analysis of *Predestination* it was inherent in the story that the central character kept meeting himself. He was frequently unaware of this, because he kept changing his appearance—but it raises the question of what happens when a time traveler meets himself. This is a trope in many time travel stories, some of which suggest that there are serious consequences. There can be, but not in the ways most stories suggest.

The earliest suggestion I can recall that meeting yourself would have dire repercussions was in the 1983 Peter Davison *Doctor Who* episode *Mawdryn Undead*. In it, two versions of Brigadier Lethbridge-Stewart, five years apart in time, wind up touching each other, releasing a huge surge of temporal energy which happens to be ideally timed to save the day. It is an interesting notion, but the question is whether that is at all likely.

In *Back to the Future Part II* several of the characters encounter their temporal duplicates.

Jennifer is picked up by the police and "returned home". Doc's concern that something horrible might happen if Jennifer meets herself (or if Marty meets himself) is patently absurd. Merely meeting yourself is not a problem. For the person from the past who meets the future self, it might be a surprise, but, like Biff, there is every chance he won't recognize himself, and even if he does that is not necessarily a complication. For the person from the future, seeing the past self will only remind him that he once saw his future self. If they meet in the past, the anomaly has already occurred, and time is already different; if they meet in the future, we are certainly on the timeline of the self who came to the future previously and returned to the past (because if not, the future self would not be there). The only danger inherent in such a meeting is that the younger self may learn something which will change his life in a way which will change the older self such that the younger self cannot learn this. It may be that Doc's concern is overblown for Marty's benefit: he does not really believe that Jennifer merely seeing herself will destroy the time-space continuum as he says, but is concerned that if in seeing her 48-year-old self, her 18-year-old self will decide not to marry Marty, it will throw the future into an infinity loop. In fact, Jennifer learns a great many things which could become information by which an infinity loop could be created.

Biff gave himself a sports almanac, and relying on it he makes himself a rich man. He murders George McFly in 1973—after Marty is born, but early enough to change Marty's life completely. Marty has spent most of his life in far-off boarding schools, and probably does not even know Jennifer or Doc. Meanwhile, Doc has been committed—the date is unclear, but it is clearly earlier than 1985, and thus before the creation of the time machine. Thus, Doc will not create the time machine, nor will he bring Marty into the future to save his kids, and so the aged Biff cannot take the time machine from them to the past. But wait! The future may be saved if Biff realizes that he must take the book to himself, and so invests in developing the time machine from the papers of the institutionalized Doctor Brown. Alas, he has no way of knowing that Doctor Brown was working on a time machine, and will not give those notes a second thought; and even if it happens that someone else develops time travel soon enough, and Biff is in a position to exploit it, his statement to Marty indicates that his older self never explained who he was, and his younger self has no clue that it was him. The triggering event—passing the book from the future to the past—sets off a chain which makes itself impossible, and we have an infinity loop—all of time comes to an end, perpetually repeating two alternate sixty-year timelines.

Doc's concern about encountering their selves in the past is overblown, although it is true that if their younger selves recognize them they will know more about the future than they otherwise did (such as that they survived and returned to the past for some reason). Also, if Marty's presence here in this sequence disrupts the events of his prior visit such that the other Marty can't return to the future, he creates an infinity loop (since he cannot then prevent himself from doing so), so it is necessary for him to prevent Biff's thugs from attacking his temporal duplicate.

In *Timecop*, the "real" danger in time travel, according to Matuzak, is the danger that the same matter might occupy the same space, and thus that a time traveler might meet his younger self. No one knows what would happen, but we see the result when time traveling Walker throws past McComb into time traveling McComb. It is not pretty, and it is definitely fatal to both of them. It makes for a dramatic climax to the story—but is it at all credible?

Let's face it: matter cannot occupy the same space at the same time. That's a rule that defines matter. If a time traveler materializes in the past, he must displace the matter that is already there—usually air, sometimes water, hopefully nothing solid. When we touch objects, they generally move and our own bodies always compress slightly, precisely because matter cannot occupy the same space as other matter. The fear here, though, seems to be that the reaction would be different if the same molecules came in contact with themselves.

That seems immediately to be a ridiculous concern. It simply could not happen. In a very material sense, you are not the same person you were ten years ago.

Any such contact, assuming it happened to get past our clothing (and seriously, how often do you wear anything you owned ten years ago?), would be epidermis to epidermis—the outer layer of skin. What distinguishes the epidermis is that it is comprised almost entirely of dead dermal tissue, cells which died to create an outer shield for the body, and that that skin is constantly wearing off and being replaced by freshly dead dermal cells. It does not take very long at all for the entire exterior of your body to be replaced completely. We notice it with the slightly different cells that comprise our hair and our nails, but it is happening with our exterior constantly: the old is wearing off and being replaced with new. When McComb touches McComb, it is not the same skin.

Beneath that, there are some parts of the body that remain mostly the same—the teeth, parts of the bones, some cellular membranes—but not only are there many cells being destroyed and replaced (red blood cells, white blood cells), the human body is over half water by weight (estimates range from 45% to 75%, depending on factors such as age and physique), and the water is constantly recycled. Given a decade, if the same water molecule is anywhere in your body, it is a remarkable coincidence. That means that at least half the mass of your body will have been replaced given a decade.

Besides, if we're talking *matter*, we have to be talking *molecules*. Thus even if some part of McComb that does not constantly regenerate—such as a tooth—touches that of his counterpart, the precision necessary for this molecule to contact this molecule is incredible. There is about as much danger in the possibility that a time traveler would drink a glass of water in which there happened to be an atom of hydrogen already present in his saliva.

Apart from all that, there is no particular reason to suppose that molecules are individualized in some way. We know that they pass electrons to each other, and thus just as your body is constantly replacing its outer shell, so too all of the atoms within it are constantly replacing theirs. The notion of "the same matter" is almost devoid of meaning on any level.

The issue is addressed again in *Mr. Peabody and Sherman*. After losing Penny in the past, Sherman returns to the present to get help from Mister Peabody. Arriving in the present, Sherman encounters himself. He should have anticipated this, but then, he's a child. The problem is that Peabody also encounters himself, and he certainly knows better than to travel to a time when he already exists—he undoubtedly wrote the rule and built the safety into the machine.

The real problem with encountering yourself is seen quite clearly here.

Older Sherman is here with older Penny as younger Sherman arrives. Younger Sherman has just come from losing younger Penny in Egypt, so she is not with him. Just as older Sherman came to enlist younger Peabody in the effort to save older Peabody, so, too, younger Sherman is arriving to enlist younger Peabody to save younger Penny. Already the schedule is off—younger Peabody and Sherman do not leave on time to save younger Penny, so older Sherman and Penny cannot have arrived here and older Peabody cannot be following from the past.

Put more simply, what older Sherman and Penny do has changed their own histories such that Sherman will never take Peabody to Egypt to rescue Penny, and ultimately Sherman and Penny will not appear here to talk to Peabody. That means that since they will never arrive here, they will not interfere with events at this moment, and so younger Sherman and Peabody will rescue Penny from Egypt and go through the events we have seen up to the moment when older Sherman and Penny arrive here to ask for help rescuing Peabody. We have an infinity loop.

As to Peabody's arrival, it makes sense this far: Peabody knows that (younger) Sherman came and took him away, and the time that has elapsed since that event is sufficient that his own younger self should have left for Egypt. He did not anticipate that Sherman would have intervened and altered history such that the younger Peabody is still present, and so it was an unpredictable problem. However, the fact is that older Sherman and Penny have already prevented the departure of younger Sherman and Peabody, and so history is trapped and cannot reach the moment of older Peabody's arrival. That moment does not exist. This is the kind of disaster that can happen when you encounter your temporal duplicate: you change your own history such that you are not who you then were and thus are not who you are. It is one of the most dangerous and complicated anomalies.

We still have the problem, though, that there are now two Peabodys and two Shermans present, and we add an angry and prejudiced Miss Grunion to the mix to get an explosive and disastrous situation.

The movie then does something unanticipated. As Miss Grunion grabs the two Shermans, they slap hands with each other and stick together. Something then seems to draw them into each other, and as the temporally duplicated Mr. Peabodys attempt to prevent this they, too, are drawn into each other such that there is only one Sherman and one Peabody.

This is not less ridiculous than the result in Timecop, in which the doppelgangers fused into a protoplasmic mass and vanished from the world. We know of no reason why touching your past or future self would be different from touching anyone, or indeed anything, else in the past or future, other than the difference created by meeting yourself and so changing your own history.

Besides, if this could happen it would lead to a far worse disaster. Let us suppose that somehow this encounter has not led to an infinity loop—perhaps younger Sherman and Peabody still have time to make their trip to the past to rescue Penny. The fact that the duplicates have now fused into a single version of each means history might never be resolved—far from preventing a paradox, this phenomenon has nearly guaranteed one. It is absolutely essential to history that Peabody and Sherman—now the only ones who exist—leave immediately to rescue Penny in the past; yet that means that when they return to the present they will again encounter themselves. This time the older Sherman might realize that he cannot touch the younger Sherman, and so they might avoid fusing; in that case, the younger Peabody and Sherman must immediately leave to rescue Penny. However, when that younger Sherman returns as the older Sherman, he does not know not to touch his younger self, and so the fusing will happen. We again have an infinity loop. That is the temporal disaster we needed to avoid, and now it seems unavoidable.

In *Looper* it is inherent to the plan that some time travelers encounter their younger selves. In the future time travel is used to send victims back to be killed by contract killers in the past, due to a problem with disposing of bodies in the future. The contract says that each victim will have payment in silver strapped to his back except the last, who will be the hitman himself and will have payment in gold. The theory is that once the hitman has killed his older self he will have thirty years to enjoy his payoff, after which he will be sent back to be executed. What makes the story interesting is that two of the older hitmen manage to escape their younger selves and run, and the film plays with the relationships between them. Much of what happens with Seth makes no sense, but is not so much about meeting yourself as about altering your past, and helps us understand what happens with Joe.

The issue with Seth is that when the older Seth runs, the crime boss captures younger Seth and begins torturing and disfiguring him. He carves a message in younger Seth's arm which appears as scar tissue on the arm of his older counterpart. He begins crippling limbs, starting with fingers, which one by one older Seth loses.

The problem here (which we saw previously in *Frequency*) is that there is no sense to limbs vanishing abruptly in the future. That older version of Seth has only two possible histories, one in which he was sent to the past intact and so has all of his limbs, and the other in which he has been crippled for the past thirty years since his younger self was tortured. He is not going to suddenly lose a body part; he is going to have his body parts intact until he is replaced by a different version of himself who lost those body parts thirty years ago and has been living without them for the past three decades.

That impacts our understanding of what happens with Joe.

When older Joe manages to disable younger Joe and escape, he leaves a message for his younger self to run for his life; he knows what happened to Seth. Older Joe's problem, really, is that he has no memories of what younger Joe is going to do, only of what he did when he was younger Joe, and he has changed that.

What the film does is have older Joe's memories update as younger Joe creates the new history. Thus the elder does not know what the younger is going to do, but instead knows what the younger has already done and what he is thinking about doing, slightly delayed. However, as it is with Seth's body parts, either our time traveler has all the memories of the next thirty years that are about to happen, or he has none of them. However, his impact on his younger self is changing who he will be in thirty years when that version of himself is sent to the past, and that version will remember all of these meetings and probably change things.

That does not happen, of course, because younger Joe decides he knows how to stop older Joe from killing the mother of the boy. It creates an infinity loop, rather blatantly, but fits with the way the film has been handling the interactions of the two iterations of the character. It would not actually work that way under any functional theory of time.

There is danger in meeting yourself in the past, but that danger lies in the fact that you alter your own history, and your younger self will now have some knowledge gained from your older self that your older self did not then have. That, though, happens simply by seeing or hearing or otherwise sensing the presence of someone who was not present in the previous history, and is complicated even when the younger self does not know that something is different from what happened to his older self. Consider when Harry Potter saves himself.[34] Those kinds of predestination paradoxes are very unstable. But despite its popularity in time travel stories, there is no reason why touching your past self should be any different from touching any other person in the past.

Jumping Into Bodies

If this was done before *Quantum Leap*, I am unaware of it.[35] In that clever series the time traveler effectively became someone in the past and lived that person's life for a while, attempting to figure out what went wrong and how to fix it.[36] It has since been used in other ways in other time travel stories, including *Source Code* and, I am informed, the television series *Dark*. Sometimes the leaper lands in his own body in the past, as in *Peggy Sue Got Married*, *Donnie Darko*, the *Butterfly Effect* series, *The Hot Tub Time Machine*, *When We First Met*, and possibly *Premonition*. In those instances the concept is

---

[34] In *Harry Potter and the Prisoner of Azkaban*.
[35] I am told that *The Dragon and the George* by Gordon R. Dickson involved the mind of a modern history professor being sent into the body of a dragon in a medieval fantasy world. The novel was published 1976, based on a short story by the same author published in September, 1957, in *The Magazine of Fantasy and Science Fiction*. It is similar in concept, although we would probably have to conclude it was dimension travel, as the existence of dragons has never been adequately proved in our universe. Thanks to Bryan Ray for calling my attention to this.
[36] The theory of the show was that the time traveler was always himself but appeared to other people (but not to animals or very young children) to be the person he replaced. It still was effectively that the time traveler replaced someone else.

that the time traveler gets to relive part of his life and change it to suit, and in that sense is a variant of Replacement Theory. A similar thing happens in time loop stories like *Groundhog Day*, *12:01*, and *Edge of Tomorrow*, as the time traveling character is reliving his own past.

The obvious difficulty with this method in many cases is in the question of what happened to the person being replaced. In *Quantum Leap* it was explained that when Sam took the place of someone in the past, that person leapt to the future and was kept in a containment. Sam is living that person's life, but that person is oblivious to what he is doing, and when he returns he will have no memory of anything that he, controlled by Sam, did during the time he was actually in the future.

This problem is even more common when time travelers return to the future after having changed the past. When Marty McFly awakens in the home of his affluent family, finding that his father's book has just been published, he seems to have replaced the version of himself who grew up in that household. Where are those memories? Many movies have this problem, including *The Hot Tub Time Machine*, *O Homem Do Futuro*, and *When We First Met*. The problem is complicated in *A Sound of Thunder*, as the time travelers return to a world in which it is doubtful that they, or anyone remotely like them, would even have existed, but remember the world that logically must never have been.

Frequency has an intriguing solution to this problem. Although John Sullivan never travels to the past, he sends information to the past which saves his father's life but costs the life of his mother. He explains to friends that he has two sets of memories, those of the world in which his father died in the fire and those in which his father went the other direction and plunged into the lake to escape the flames. Yet at what point in time did he get those dual memories? After all, once his father escapes the fire there is no history in which John grew up without him, and so no John who lived that life. Does the teenaged John wonder why part of him thinks his father died? Does he sometimes get confused about where he is or what he is doing because he is simultaneously recalling what he was doing in the original history at this moment? Does this schizophrenic dual reality cause him to need therapy? While it's a clever notion, it only works if we somehow assume that both histories happened and the one John lived through each of them, perhaps sequentially, perhaps simultaneously. Having two sets of memories of the same period of history does not seem psychologically plausible unless the one experiencing it has the sense of having lived them in sequence. In addition, if the time traveler changed history by leaping into the body of someone in the past and he then does not leave from the future to leap into that body in the past, then he does not arrive in the past and we have an infinity loop. Further, once history has been changed, it should not be possible for Al to access the original version. It is still necessary to preserve the altered history if history has been altered.

In *Source Code*, the traveler leaps into the body of someone who is about to die, on the assumption that nothing will change that. Unfortunately, though, in at least one iteration of history the time traveler does change that, preventing the death of the person whose body he borrowed. Then complicating it further he stays in that body for that history, taking over the life of the man who now did not die. How does that work? He knows nothing about this person—he won't recognize his supposed mother's voice if she calls him on the phone, won't know where he works or what he does, won't remember any of his friends or anything they do when they get together. Even had he done extensive research into the life of the person he became, there would be significant gaps in his knowledge—not to mention that he would have undergone a complete personality shift that would undoubtedly be noticed by someone.

Apart from that, leaping into bodies faces all the other problems associated with time travel: if you change the past based on knowledge of the future, you erase that knowledge and so will not know to change the past. You would need some kind of omniscience to do that.

How to Change the Past

Perennially readers write to ask whether this or that scheme would work to change the past. It usually involves telling your younger self to do what you have done. He is not you, though. Even apart from the fact that you did not have the experience of being told what to do, he lived in a world that you made, not the world you experienced. He does not have the same motivation, which will show even when he in turn tries to tell his younger self what to do. The best hope is that the sawtooth snap this creates ultimately resolves to a stable history and an N-jump; the probabilities are against it.

Yet there might be a way to control the variables and hopefully create a successful outcome. If you have seen the show *7 Days*, they had part of it, but not near enough. We will use them as the model, though, to finish the system.

To do this you need four teams and a time machine capable of short hops of a precise and unvarying length. Seven days is the length we will use. Our teams are the investigators, the time travelers, the responders, and the coordinators. Each has an essential part in the plan.

When something happens that needs to be prevented after the fact, the investigators learn everything they can about it. The quicker they can compile their information, the more time will be available at the other end to prevent it. Everything they learn they save in something akin to a digital form, and give to the coordinators.

The coordinators work with all three of the other teams; otherwise those teams have no contact with each other. At this point, the coordinators receive the information from the investigators and store it on a durable permanent medium such as a digital video disk. They have a stock of these, and they use the first one in the stock. That matters. They have a delivery system that will transfer that disk to the time travel team without any other contact. The delivery time is encoded on the disk, and the delivery system delivers it at the designated time.

The time travel team is completely isolated from the outside world. All information coming into their base is on a seven-day delay. They do not know what happened today, or yesterday, or any day back seven days. That means they do not know whether the events on the disk have actually happened or have already been prevented. As far as they know, this is something that really did happen in the seven days which are blacked out in their news feeds. Any member of the team who leaves the enclave is quarantined for seven days upon his return so as not to contaminate the information base inside the enclave. When they receive the disk, they prepare to send it back in time seven days. If they send someone with it, it is always the same person, and he remains in quarantine when he returns so as not to contaminate their news feed. Note that he must not re-enter the enclave until after his doppelganger has departed.

The time traveler himself delivers the data disk to the coordinating team, and then goes into quarantine. The coordinating team now does two things. First, they take the first disk from their supply—which is the temporal duplicate of the disk they just received, the same disk seven days younger—and they copy the data from the disk they received to the new disk. That disk will include the encoding for when it is to be delivered, so they prepare to deliver it to the time travel team. The data they received is also given to the response team, which uses the received information to attempt to prevent the disaster. If they succeed, the coordinators inform the time travel team by means of their seven-day delay.

Note that the time travel team does not know when they receive the disk whether they have already succeeded in preventing the disaster recorded; they can only assume that it needs to be prevented. Likewise, the response team does not know whether they have already prevented this in a previous timeline or how they did it, and can only do what they think would be best. Since all the information is the same, it is for everyone the first time through every time, and so history stabilizes into the corrected form, and we have changed history.

Let's steal a plot from *Déjà Vu*, but change their time travel methodology a bit. They had a wormhole whose past end was four days, six hours, three minutes, forty-five seconds, and fourteen-point-five nanoseconds in the past. They set it so that they could view events in the past by somehow manipulating the position of the past end. Let us for convenience suggest that their wormhole is exactly seven days—that is, 168 hours, or 10,080 minutes, or 604,800 seconds—in the past, and that rather than seeing through it they use it to send a traveler back that temporal distance. They have the four teams set up as just described.

In this story, someone has planted a bomb on a ferry as an attack on the crew of the U.S.S. Nimitz which is holding a scheduled shore leave party on the boat. The bomb is in a stolen sport-utility vehicle whose owner was murdered earlier in the day, and the exploding vehicle caused the fuel tanks to explode, destroying the ship and killing most of those aboard. FBI, DEA, NSA, and ATF all respond and begin investigating, but our time travel investigation team quickly steps in and takes control of this. Using more traditional means, they determine that Carroll Oerstadt murdered Clair Kuchever and stole her vehicle, then drove it onto the ferry in the middle of the night before the bombing. They put together all this information and give it to the coordinating team, who compiles it on a disk with a delivery time stamp and puts it in the delivery system which transfers it to the time travel team.

The time travel team gives the disk to their time traveler. They might review it before they do so, but what matters is that they are sending the disk back seven days in the care of someone who will ensure that it is delivered to the coordinating team seven days in the past. Hopefully, if the investigators managed to solve the problem quickly enough, this is three or four days before the bombing happens. The coordinating team now copies the disk and puts the copy in the delivery system, which will read the delivery time stamp and so deliver the new disk to the time travel team at the same time that they would have received the original disk. Meanwhile, the original disk is given to the response team.

Reviewing the data, the response team determines that Carroll Oerstadt is responsible for the bombing, and perhaps a day prior to the explosion he stole a car and killed its owner, whom they have identified, and so they stake out the home of Clair Kuchever and detain Oerstadt.

The difficulty here, as in *Minority Report*, is that they have prevented the crime and will have great difficulty proving in a court of law that Oerstadt is actually guilty of anything. However, they have prevented a major disaster and identified a bomber whom they will undoubtedly watch in the future.

Meanwhile, the time travel team does not know that the bombing has been prevented, so they proceed to deliver the disk to the coordinating team seven days in the past, which, because they are the same people doing the same thing, will happen at the same time, and this version of history will be identical to the one it replaces, with the coordinating team again delivering a new copy of the data to the time travel team and providing the old one to the response team who will again apprehend Oerstadt and prevent the disaster. We have created an N-jump, history resolving into the altered version. We have in fact successfully changed the past, and kept it in its changed form.

We cannot stretch the time too long. Quite apart from the fact that it would be difficult to isolate anyone for a significantly longer time, we also risk overlapping disasters. But for something like the Backstep Program, this would be a workable system, and nearly foolproof, depending on the competence of your local fools.

Foreseeing the Future

There is a trope related to time travel, sometimes actually being time travel and sometimes not. These are stories in which someone gets information about the future to change the past without actually traveling through time. A few films in which someone changes the future based on knowledge of the future without actually traveling through time include *Frequency*, *Minority Report*, *Next*, *Watchmen*, *X-Men: Days of Future Past*, *Time Lapse*, and *Mirage*. Looking at just those films, we find three distinct concepts with entirely different consequences.

The theory in *Watchmen* is very difficult and not terribly useful outside its own parameters. The character Dr. Manhattan has been the victim of a nuclear accident which has in some sense disconnected him from normal spacetime. For him all times are now and all futures are real, and so he knows what might happen in the sense that he knows all the possible futures and knows that one of them will happen to the people around him. It is very problematic, reminiscent of the prescience of Maud'Dib, who could see all possible futures and make choices to bring about the one he preferred,[37] except that for Dr. Manhattan all those futures happen and it's only a matter of which one he chooses to experience.

For our other stories, we have two distinct types of foreknowledge.

Frequency presented a story in which John Sullivan, in 1999, starts talking to his father Frank Sullivan in 1969. This is clearly information traveling from the future to the past, and based on it Frank changes his actions and so

---

[37] In the book trilogy *Dune* by Frank Herbert.

escapes death in a fire that year. The film *Mirage* similarly has a conversation, this time between strangers, a boy in 1985 and a woman in 2010, in which the boy is a witness to a murder but no one believes him until the woman proves it. In these cases, information about the future is being transmitted to the past, and people in the past are acting on it in ways that change the future.

We have a similar situation in *X-Men: Days of Future Past*. This, though, is closer to a body jumping story. A hero named Kitty has the power to enable someone else to take over his own body in the past and so deliver a message to the heroes that they are in danger and need to move. Still, this is a situation in which real information about the future is delivered to the past, and those in the past act on it to alter that future.

Let's consider this under our three fundamental theories of time.

It is obvious that the entire point is to change the future based on knowledge from the future, and as we observed about Fixed Time Theory it is not merely that you cannot change the past, you cannot change the future, either. But if you suppose that there is a difference between changing the future and changing the past, let's observe that in all three of these films we are observing a significant part of the story from the perspective of the persons in the future who are sending the information to the past, and thus effectively changing the past. If information is coming from the future, then the future already exists; and if it exists, in Fixed Time, it is fixed and cannot be altered.

We could envision this happening as a Multiple Dimension Theory story. Let us suppose that John is talking to the Frank of another universe. John saves that Frank's life—but his own history does not change, his father Frank still died in that fire. He can continue talking to him on the radio, but he has to recognize that this is not his real father, but a different father. That's not the way the stories play in the movies—the people in the future have altered their own pasts. We could have such a story, but it wouldn't be a time travel story. Further, once John has altered Frank's future, the history of Frank's world will be forever out of synch with that of John's.

So what happens if we do this under Replacement Theory?

Frank died in the fire. John grew up without Frank. Then John contacts Frank in the past and warns him that he will die in the fire if he does what he thinks is right. Frank now changes what he does, and escapes the fire alive. John grows up with his father Frank. Then John accidentally contacts Frank—but this John knows nothing of Frank dying in the fire, and so he doesn't tell him anything. Not having that warning, Frank dies in the fire, restoring the original history; we have an infinity loop, in which Frank is alive in one history and dead in the other.

If we use Dr. Manhattan's reality, Frank is essentially a human version of Schrödinger's Cat, both alive and dead. The problem remains, though, because to John Frank must be one or the other. Dr. Manhattan might be able to live in multiple realities, but the rest of us have to exist with a single set of facts.

Changing the future based on knowledge from the future is a disaster, because as soon as you have made the change you have eliminated the knowledge on which the change was based and so undone the change.

Yet there is another way to change the future based on knowledge of the future, and we see it, if not clearly at least certainly, in *Minority Report*. In that story there are three psychics each of whom is sensitive to highly emotional future events such as murders, and their mental images are transferred via mind/machine interface to a computer. A special police unit then arrests those responsible for committing murders that have not yet been committed. The peculiarity in the system is what gives the story its name: sometimes one of the three psychics, Agatha, sees a different future from the other two, known only as the twins. When she does, the police ignore and indeed erase her vision (it would otherwise be exculpatory evidence at trial), and act on the assumption that the vision of the other two is the actual future history without trying to explain why these alternate versions of those events exist. The best answer, though, is that the psychics are not seeing information from the future, but are doing something which is in a sense quite ordinary in modern cybernetics translated to a psionic ability.

We today use computers to predict the future. The two most common uses are weather forecasting and economic forecasting. They both work in similar ways: we provide the computer with massive amounts of data, so that it has every fact we think is relevant to the present situation and a program defining how these facts interact, and then we have the computer calculate what those values will be in an hour, a day, a week, a month, a year. The nearer the projection the more accurate it will be, such that tomorrow's weather forecast is more likely to be correct than that for next week, and the economic prediction for next month is less certain than that for next week. This is in large part because there are likely to be factors of which we were unaware that impacted the results. After all, you can line up a pool shot perfectly, but if at the moment you hit the cue there is a small earth tremor it will mess up the shot. Such predictive programs are imperfect, but they are usually rather accurate.

In the same way, we can imagine that Agatha and the twins are each collecting and processing information about current events on a grand scale, such that they subconsciously know the thoughts and feelings of billions of individuals and from that can extrapolate probable actions in the near-term future. Just as many of us can predict actions and reactions of our close family members, what will make them angry, what will make them happy, what will surprise them or wear them out or touch their sense of obligation or fair play, our psychics can do this on this massive scale. They in that sense don't see anything that happens in any future history of any world (unlike Dr. Manhattan), but rather see events that they have extrapolated as the most probable futures based on information in the present.

Thus the reason that sometimes Agatha's vision differs from that of the twins is that she is gathering different information and so predicting different outcomes. No information comes from the future, so the future isn't being changed. With an economic forecast one might change the current interest rates or tighten the money supply or otherwise change facts in the present to attempt to avoid anticipated problems in the future. In the same way, Agatha and the twins are predicting the most probable future based on their extensive knowledge of the present, and the police are not changing the future but acting in the present to alter the situation and prevent the predicted future crime.

We have the same explanation for Frank Cadillac's knowledge of the future in *Next:* he psychically gathers massive quantities of information about his present situation and instantly extrapolates what will happen around him in the next two minutes. He is then able by changing his intended actions to predict what will happen instead, and so choose a course of action that will achieve an outcome he prefers. We are again reminded of Maud' Dib, who chooses what he will do with a view to steering history in the direction he desires.

It is thus possible to have precognitive abilities that are not time travel, and in that case it becomes possible to alter the predicted future without erasing the information: at the moment the prediction was made, this was the most probable outcome, and if the prediction itself becomes a factor that alters that future it is not a factor that existed at the time the prediction was made. Actually seeing the future and then changing it is a temporal disaster, but predictive abilities don't have to do that.

Toward Two-Dimensional Time

Many have written to me mentioning Poul Anderson's *Time Patrol* stories, and I have written back to say first that I had not read them and second that I did not do analyses of books for a variety of reasons. However, in 2008 one of my readers decided to mail me a copy of the complete collected stories, along with some other science fiction and fantasy books he thought would round out my familiarity with the literary side of the genre (and I did and still do thank him for that).[38] I read it, and would say that to some degree I enjoyed it; they are good stories well told. They do not, however, fit into any model of time travel familiar to me.

This led me to wonder whether there was a plausible model of time travel I had missed. I long pondered what sort of description of time might make Anderson's stories plausible, and began to tinker with a model I am presenting here. I still will not analyze books, and am not going to do so here. I merely wondered whether the *Time Patrol* stories might become possible with a different model, whether Anderson actually had a clear, coherent, and plausible theory of time and time travel from which he was working, and whether I could discover it. This is not, then, an analysis of those stories, but only an effort to develop an alternate conception of time in which stories like those, if not those stories, might be conceptually possible.[39]

---

[38] Thank you, Eric Ashley.
[39] This idea was originally published at the Gaming Outpost web site under the title *A Draft: Toward Two-Dimensional Time*, and there were some responses to it there which may have influenced this version.

In Anderson's world, people travel to the past all the time—but most of them do so because they work for an organization dedicated to preventing changes to the past. It seems that the day people, somewhere in our future, discovered time travel, they were visited by people from a yet much more distant future, an incomprehensibly distant future, who had a vested interest in preventing change to the past, who informed those earliest inventors of the technology that they were now drafted into a temporal police force to prevent anyone else from using their technology to change the world. They also recruited at least a few people from earlier times, including several from the twentieth century, to work as researchers, historians, and enforcers. These were provided with the equipment needed to travel through time, and given life extension therapy so that others from their own age would not wonder either why they aged so quickly or where they were all the time.

Even a casual fan of temporal theory should recognize that such a scenario is not possible under any of the familiar models of time. Bear with a brief exposition of the flaws.

Under Fixed Time Theory the people in the future are wasting their efforts trying to preserve the past, because the past cannot be changed—all effects of all time travel events are already part of history, and those who will at some point in the future travel to the past in some sense have already done so, have already arrived in the past. The time patrol itself is nonsense, as it is enforcing rules that cannot be broken by attempting to break them. One might as well organize a police force to enforce gravity, and say that in their efforts to enforce the law of gravity they are permitted to break it. Of course, Anderson's stories would be rather boring in that case. That's not to say that you can't tell an interesting story in fixed time—only that a story about a police force that attempts to correct changes made to history before they become serious problems is not such a story.

Parallel and Divergent Dimension Theories are vexed by the problems outlined in our earlier discussion of the two brothers. Notably, our future society that is attempting to preserve itself is simply creating other universes. As with fixed time, the society of the future cannot be changed—in this case, because its history is fixed, and the time traveler is tampering with someone else's history. Further, if we assume that someone from the future has traveled to the past and created a new universe in which that future society does not exist, then someone else from that future "fixes" the universe such that that society is restored, we have gone from having one universe in which that future society exists to having three universes—one in which the future society existed and was never endangered, one in which the future society never existed and never would have come into existence, and one in which the future society came into existence because after a traveler from the future tampered with history, someone else tampered with history again.

Even more problematic for telling such tales under the Parallel/Divergent Dimensions Theories is the problem created by the linear nature of each such dimension. Although *Back to the Future Part II* is a disaster of a time travel story, it gets this part right: once Doc and Marty are in a divergent dimension, they cannot go forward to the point from which the time traveler departed to correct the problem, but only backward to the root where the change occurred. Once a change has been made in the past that would destroy the society of the future, there is no society in that future who can detect and correct this. In that dimension, that society never came into existence.

Replacement Theory also rejects such a concept. For there to be any future universe, all anomalies must already have resolved. Let us suppose that the future society lives in twenty thousand A.D.; let us suppose that time travel is discovered in three thousand A.D. If our time travelers in three thousand travel to two thousand and completely alter history between two thousand and three thousand, maybe they will destroy time, and maybe they will be very lucky and preserve time—but by twenty thousand, that's all ancient history, more ancient to them than the earliest Egyptian inscriptions are to us. Whatever that distant future society is, it became that because of whatever changes were made to history by all time travelers before then. Their intervention cannot prevent settled history from changing, because for them the changed version is and has always been the settled version. They exist because of every change that was made, not in spite of these. There is, again, nothing to protect.

Yet the stories seem plausible. This suggests that there is a way of understanding time that does not fit any of these models.

The solution that seems to be in view is to perceive time in two dimensions.

On one level, this is similar to the conception of time outlined in our spreadsheet analogy above—similar enough that the reader should understand that concept before attempting to grasp this one. In short, there is a sense in which all of history occurs metaphysically simultaneously. The future is not unformed, merely undiscovered; the past is not unalterable, merely known. If there is a change in 2000, there is a consequent change in 3000, an immediate change because the events in 3000 are in direct causal dependency upon the events of 2000. It does not take a thousand years for the change to occur; it takes a thousand years for us to reach the place in time where we perceive what change occurred.

Each of the popular theories of time travel treats this problem differently, but in each case it can be comprehended by visualizing time as if it were space. In the fixed time theory, that space is unidimensional: time exists in a continuous line from the past to the future, and cannot be altered. With the parallel and divergent dimension theories, there are multiple timelines lying alongside each other, and travelers leaving one arrive in another. This is the concept of sideways time, suggested in a John Pertwee *Dr. Who* episode[40] and exploited in *Sliders*. In this conception, these parallel or divergent universes exist temporally "alongside" each other, but are disconnected save by the acts of time travelers, who are really dimension hoppers. Also, the past is immutable, but the future is being created.

The replacement theory gives the impression of two-dimensional time, but it does not support the concept. Rather, as with fixed time, it maintains that ultimately there is only one history of the world—but that it might be altered, and if it is then there is still only one history of the world, the other having existed only in some metaphysical past, something like the program on your video recorder that you erased to record another. Still, the description of history in this case involves tracing causal lines to determine whether the past still supports the future which supports the past. The past can be changed, but once it is, nothing from the original past remains.

---

[40] *Inferno*, season 7, 1970.

For *Time Patrol* to work, there needs to be two-dimensional time. Time moves from the past to the future, as is familiar to all of us; but it also moves laterally, from one version of history to another. Yet lateral versions of history still exist, and anything which originated in one can continue to exist in another.

What ought to make this work is the interaction between chronological time and lateral time. Having accepted that chronological history all exists simultaneously, and that time is merely the way we experience it, we can understand that a change in 2000 could cause an immediate change in twenty thousand. We would have to wait until we reached twenty thousand to see that change, but there are people in twenty thousand whose lives would change immediately.

The problem is, no one in twenty thousand would know that history had changed. It would always have been thus—unless there is another aspect to time.

That other aspect would be the lateral. This rests on the notion that change requires a medium within which to occur. In history, change occurs through chronological time: I drop my pen *now*, and it hits the floor *now*. We are accepting that this is perception, that the falling and landing of the pen are accomplished instantaneously at various points in time. Yet if we involve time travel, and we eliminate at least some of the conception of time as the medium in which changes occur (because we are now looking at global changes, changes which occur instantly across all of time), it might be asked how that change can happen. One possible answer is lateral time. With each tick of lateral time, the history of the universe is established; if a time traveler from a previous tick makes a change to the past of this one, the entire history of this tick forms to accommodate that change, but still accommodates causes which have moved from the previous tick into this one.

That means that time is not really moving "forward" toward the end of history, but it is really moving "sideways" from iteration to iteration of all of history. Anyone who moves backward in time also moves sideways, into a subsequent iteration of the entire history of the world. He might undo his own birth, but this paradox is inconsequential, since his birth still exists in the iteration of history from which he came, a spatio-temporal location as real yet inaccessible to him as yesterday is to us.

Unfortunately, this does not solve the anomalies portrayed in Anderson's work. As noted in connection with parallel and divergent dimension theory, the problem still remains that in any iteration of history in which the desired future society is destroyed, it does not exist. Thus it cannot send enforcers back to correct a problem that it cannot detect.

Anderson's answer to this seems plausible on its face. Time Patrol training facilities exist in an era in the distant past—prehistoric past, dinosaur past. This facility includes among its resources everything that is known about the future history of the world, and specifically what events are thought to be important in establishing the existence of the future society which sponsors their work. Since this facility exists at a point prior to any known ventures to the past, it continues to exist despite changes made in its future. Since it can still possess information and personnel and equipment which were delivered from the future of one iteration of time, the shifts in the existence of the future society will not impact its ability to perform its mission.

The problem is finding the relationship between historic time and lateral time. Think of it this way: those who are stationed in the past are charged with preserving the future whose records they have in their files. The moment some change in that future is detected, one of their own must undertake to identify the change, determine how to restore what would be as near as possible to the original history, and travel to whatever point in time this repair can be implemented. But it is still that aspect of knowing when the future has been changed that creates the problem.

From the perspective of historic time, the answer seems to be either that it will be changed or it will not be changed at a particular moment. Our theory of two-dimensional time, though, suggests that there is an original history of the universe in which no time travel events ever occurred, and then "when" someone traveled to the past, we moved laterally across time to the altered history. That lateral shift, though, is not something we can perceive or experience; it also is not something that happens at a specific moment in the historic timeline: it happens to all of history simultaneously. Thus at time L1 (L for Lateral), the original history of the world, no one can detect any change in history. Then at time L2, following the "first" time travel event, everyone who can detect such events detects that one simultaneously.

To clarify, let us suppose that the time patrol base in the past exists for a thousand years. In the L1 moment, no one ever detected any change in the history of the world. Then as we moved to the L2 moment, a change occurred. That change became detectable—but it was detectable simultaneously through every second of those thousand years. Then the L3 moment occurs, due to another trip through time, and both changes are detectable for that entire thousand years. More significantly, those who live in the time patrol base when it reaches L3 cannot know which of the two detected anomalies occurred first. Further, those who live at year 1 have no means of knowing whether they are actually detecting a change made by their own people at year 5. In fact, every time someone is sent back to the base, it creates a detectable anomaly, a change in the history of the world, a next tick on the lateral timeline. Without a complete roster of everyone who was recruited and sent for training, the people at year 1 cannot sort out who is what.

You might think it a simple enough matter for the people at year 1000 to send that complete roster to the people at year 1, so that they can use that to compare against whatever they detect. The problem is, as time moves laterally the information at year 1000 will change but the information already sent to year 1 will not—otherwise, the record of the history of the future society these people are trying to preserve would also change, and they could not know whether history had changed or not, nor what it was they were trying to preserve. In order for the people at year 1 to have current information, the people at year 1000 must send back an update with every click of lateral time. That, though, is impossible. Even were we to suppose that our time travelers have a way of detecting clicks of lateral time, they cannot act in lateral time, only in linear time. They can send a roster back from the earliest lateral moment in which they have that roster, but in order to send an updated roster they would have to be able to send both rosters simultaneously. Again, if we assume that having sent the one roster at L1 they automatically will have sent the most currently updated roster at L1000, we must also assume that whatever history of the world was sent back from the distant future is that version that exists at L1000, having been updated by virtue of the fact that the people in that future will have sent the most current version.

Too, this ignores the detail that the very sending of the current roster is itself an event, a change to history which needs to be recorded on the roster of those events which are to be ignored.

The concept of lateral time does not solve the problems created by the *Time Patrol* stories, nor does it appear to resolve any other problems. My impression is that Replacement Theory is still the best theory. At the same time, the concept of lateral time might help resolve some of the uncertainties of two-dimensional time, particularly in relation to what is the future for a time traveler who travels to a point prior to a previous anomaly. In any case, I offer this as a starting point for any readers who think there might be something of value here, and invite you to make your suggestions for how this might be brought to a functional theory of time.

The Perpetual Barbecue

This story was originally published in RPG Review, *in September 1999; it has since been republished in* Multiverser: The Second Book of Worlds *from Valdron Inc, and on the Temporal Anomalies in Popular Time Travel Movies web site. It has been reprinted here, with some minor edits.*

It was the worst Unification Day picnic in memory. Although it was a very promising day in August, with blue skies lightly marbled with wisps of white, and temperatures coming into the low thirties—hot enough for swimming, but not too hot for volleyball and softball—things just went horribly wrong. Certainly Lakeside Park was ready: the softball field had been mowed and lined, the volleyball and tennis nets were up, and fireworks were set up in one corner of the field which had been roped off and was being carefully watched by a few pyrotechnics experts. But somehow things started going wrong, terribly wrong. As Mrs. Roddenfield said, it was almost as if some vital piece of the puzzle was missing, and without it the puzzle made no sense at all.

It started when the fireworks company arrived, and found Dr. Arnold Hess down by the lake. He had apparently come down in the middle of the night, and had a heart attack beside the water. But it was agreed that news of this would ruin the day, so the ambulance quietly carried his body away, and no one was told.

Then at about eleven in the morning Johnny Adams, showing off to Myra Wilson, had popped the clutch on his reconditioned '98 Camaro, and crashed right into Pete Thompson, who happened at that moment to be crossing the driveway to get to the baseball game. Of course, the ambulance was at the picnic—the whole town was expected to be there, so it seemed prudent to be prepared—but word from the hospital wasn't promising late in the day.

Still, Unification Day was the kind of holiday people celebrated. Many of the adults could remember the tensions of the late twentieth century, when the cold war between the superpowers gave way to the third-world terrorist assault. Bringing the world into a single economic and political system had been difficult—it had taken a generation to accomplish—but it had brought about peace and a certain level of prosperity. The world wasn't really unified; it would take a few more generations before people all over the globe regarded each other as countrymen, as it had for the United States to perceive itself as one nation instead of many, as it had for Americans of every color to see each other as Americans without color. But it had begun, and every day brought the world closer together. It was a day to celebrate. A serious accident of this sort would put a damper on everyone's spirits, but accidents did happen, and the celebration would continue.

And everything else went smoothly for the next few hours. Of course, Mrs. Rogers forgot to bring her potato salad, and a lot of people were disappointed, because it was usually very good. But the softball game was going as scheduled, and there had been several volleyball matches. Everyone had brought their own picnic lunches, and already the smell of barbecued chicken and ribs was in the air: Pete's Barbecue Shack had agreed to feed the entire town, and bill the township for costs only.

But about two in the afternoon, some kids were crying from the woods. At first, there was little attention to this; kids cry at picnics. A few parents looked up to see if their child was among the injured. But the noise grew, and people started moving toward it as the first of the children came from the woods covered with yellow jackets. They had been playing tag, and had run right through a nest of the nasty wasps. Half a dozen children had been severely stung, and as no one had brought baking soda, several families packed up to leave. But the worst was Michelle Potts. She was allergic to bees—and since Dan Jackson of the ambulance crew was Pete Thompson's half-brother, the ambulance had stayed a while at the hospital over in Greenwich. Jim, the lifeguard, came running from the beach when he heard; there was little in his first aid kit that would help, but at least he was trained in basic life support. He spent most of an hour keeping the little girl alive; she might survive—the ambulance returned while there was still hope, and rushed her to Greenwich as quickly as they could turn around.

But while Jim was taking care of Michelle, Bob Walker drowned. He had had a bit too much beer, and stumbled on the dock when he went down looking for his wife. With everyone back up watching Jim with Michelle, no one saw him fall into the water or heard his feeble cries for help. He was found floating in the lake when everyone returned; it was too late.

It was about this time that Mrs. Roddenfield was talking with Mrs. Kowalski. "It's been a horrible day," said Mrs. Kowalski. "I don't think I'll ever have quite the same feeling about Unification Day again. As long as I live, I'll remember this day."

"I know what you mean. Mrs. Thompson is a wreck; Bob and Mary Potts—I don't think I've ever seen anyone so upset. And the Walkers! There's a family in shock. It's just like a puzzle with a piece missing, and you can't make sense of anything because nothing fits without that piece. Know what I mean?"

But Mrs. Kowalski really had no idea what Mrs. Roddenfield meant; perhaps Mrs. Roddenfield didn't really know, either.

Still, many people stayed, partly because of the promise of chicken and ribs, partly because they were encouraged by Councilman Jake Jones, who said that the town would have to pay for the food even if nobody ate it. Besides, it was Unification Day; whatever happens, the world should celebrate its unity.

So there were still a few hundred people around when Bill Peterson's fishing line snagged something, and out of the lake came the body of Dr. Arnold Hess. The elderly scientist, who had taught and researched physics for years before his retirement a few years back, had been fatally shot at close range, and his body rather ineptly weighted and tossed in the lake. Members of the fireworks crew seemed especially bothered by this; one asked if Dr. Hess had a twin brother—but no one mentioned having found the same man dead on that same beach that morning. After this, the evening fireworks were canceled, and everyone who remained went home.

Dr. Hess, of course, did not go home. His house stood empty and dark. The strange equipment in his basement was quiet, and the notes on his desk lay open and unfinished. In an open drawer was an open and empty gun case and an open box of ammunition.

And during this night, while all were asleep save a few who watch the night, something happened to time itself, something which must have happened before, but of which no one was aware.

As day dawned, a lone figure stood on the hill at Lakeside Park. Dr. Arnold Hess was looking forward to Unification Day; he wanted it to be the perfect picnic. He wanted it to be the best day it could be. His hopes appeared to be arriving—the day dawned clear and bright; the morning dew quickly evaporated in the warming sun and soft breeze. Soon picnickers would join him. The pyrotechnics experts began to set up fireworks.

It was important that he not interfere; the day must go perfectly without his involvement. He would watch; he would wander among the picnickers and enjoy the celebration.

About eleven o'clock, as he was walking over to see if Mrs. Rogers had brought her famous potato salad, Pete Thompson stopped him for a word. Pete still taught up at the university, and sometimes shared the latest gossip, which he preferred to call news. He also asked how Dr. Hess was and whether he was still working on anything. They were interrupted by the squeal of tires, as Johnny Adams peeled out of the parking lot with Myra Wilson on the seat next to him. "Kids!" muttered Pete. "One day that one's going to hit something, and it won't be pretty."

"Yes," Dr. Hess agreed, "given enough time, it's bound to happen. But perhaps he'll learn before it's too late."

"I hope so. I'd hate to see a good kid like that ruin his life over something stupid. Well, I'd better get over to the game; I promised to pitch the third inning. Come by for a burger later, and we'll catch up."

Dr. Hess continued over to see Mrs. Rogers. She had forgotten to bring potato salad; but he suggested that if she asked around, it might not be too late to do something about that. She got her kids scrounging around, and soon had everything she needed. The Simonsons had brought several sacks of potatoes for shish-kabob, and Mr. Cyminski had more onion and mayonnaise than his hamburgers could hold, and the Bahts were having salad from which they could spare some celery and vinegar, so with the help of a pot that Richard Flannagan had brought to boil corn, the potato salad was made, and everyone who contributed to it enjoyed it immensely.

As for Dr. Hess, he moved on to see Mrs. Roddenfield, who was making peanut butter and apple jelly sandwiches for her kids, just to hold them until the barbecued chicken and ribs were ready for dinner. There were a couple of bugs buzzing around the jelly.

"Bees," he stated flatly.

"Yellow jackets," Mrs. Roddenfield replied. "There must be a nest off in the woods; that seems to be where they're coming from. I've told my kids to stay out of there. They can be really nasty. Can I fix you a sandwich, Doctor?"

"No thanks, Margaret. I've been promised some potato salad when it's ready."

"Suit yourself. Angie does make good potato salad, but I'll be surprised if it comes out so good this time as it does when she makes it at home. What about you, Mr. Potts? Maybe Michelle would like one?"

Dr. Hess moved off to look at the fireworks rig. He spent some time here, just looking at the intricate wiring and blocks of explosive powder, all connected to a computer which would match the firing sequence to the music. The celebration would end well, as a night to remember.

When he came back for his potato salad, Mr. Potts was there again. "Michelle," he called out to his daughter, "You'd better stay over here, honey. Maybe you other kids should stay here, too—there's a wasps' nest in the woods, and you don't want to get stung. Why don't you try to get a volleyball game going instead?" The kids seemed to agree that they did not want to get stung, and turned their attention to the playground equipment.

After lunch, while the smell of chicken and ribs and hickory smoke filled the air, Bob Walker went down to the waterfront to look for his wife. He was a bit unsteady from the beer, and slipped on the wet dock. Jim immediately pulled him back out and scolded him for being so careless; but the laughter from several onlookers probably had more sting than the lifeguard's good-natured rebuke.

Mrs. Kowalski commented to Mrs. Roddenfield that it had been a wonderful picnic—the perfect Unification Day celebration. Mrs. Roddenfield agreed, saying that wherever President Mlambo was celebrating, even he couldn't be enjoying himself more.

The chicken and ribs were delicious, and everyone ate heartily. Bill Peterson caught a large bass, and Pete said his barbecue people would clean it and cook it right up for him on the hickory fire, such that he would never have had such a fish before or ever again after. Bill thought it was the best fish he'd ever eaten, and several who got a taste of it agreed.

The fireworks that night may have been better in Peking or Washington or Moscow or Paris; but nowhere were they more enjoyed. They began about nine, as things really were getting dark, and kept the sky lit for most of an hour. The computer followed its program perfectly, keeping the explosions in synch with the collection of music which spanned several centuries and came from every continent. If there was a message in it, it was that unification was a good thing which had brought only good to the world.

But if there was such a message, most of the crowd was too tired to know it. The park cleared within half an hour, and but for a few who watch the world in the night, everyone returned home to collapse. When tomorrow dawned, they would have to return to work, school, and the troubles of daily life; but today had been a celebration to remember.

Dr. Arnold Hess soon found himself alone in the dark. He walked home. It had been a short distance, and he had not wanted his car to be in the parking lot all day. He turned on the radio when he stepped inside; he should listen to the news, just to be certain that today didn't include the kind of disaster no one wants to remember. The news was filled with reports of Unification day celebrations from around the world. It was a good day everywhere. He was satisfied; it had been perfect.

Down in the basement, he started up the equipment which he had kept so secret. He made a few final notes, and carefully closed everything. Removing the pistol from its case in his desk drawer, he stepped into the machine; and that which had happened unbidden to time at least once before was this time caused to happen for what to anyone's knowledge might have been the first time or the hundred first.

As Dr. Arnold Hess stood in the pre-dawn darkness at Lakeside Park, a lone figure approached him. He quickly recognized it; it was he.

"It works, then."

"Apparently so. Time travel is now possible. You know what we have to do."

"Couldn't we just hide everything—destroy the notes, dismantle the machine, forget we ever knew anything about time?"

"You know science doesn't work that way. When something is ripe for discovery, there's no stopping it. We might delay it for a few years, but soon enough someone will discover it as we have—and once it's been discovered, a disaster is inevitable."

"You're right, of course. Even if no one gets the crazy idea of trying to change history, eventually someone will do something which will create an infinity loop, and time will be caught forever repeating the same two alternate histories."

"Which is why we have to do it first. The world could be caught in a time loop of which one history is the wars and destruction of the twentieth and early twenty-first centuries..."

"...and the other, a far worse scenario in which unification never happens, and humanity destroys itself. But is that really our problem?"

"I'm taking responsibility for what I've created. The day ahead is the perfect day—everyone we know and love will have a wonderful time at the picnic, and we will assure that for all eternity all they will know is this celebration."

"And you've done nothing which will be missed?"

"I will stay here, and repeat my actions. They are of no consequence, I'm sure. I asked a few people how the day was going, and ate some potato salad."

"O.K., then. I could wish I were on the other side of the loop, but there's no way to change that. Let's get it over with."

And those were the last words that Dr. Arnold Hess ever spoke; for at that moment, he shot himself—although whether it could be called suicide or murder would have baffled jurisprudence for generations, were there to be any future generations. After having shot himself and convinced himself that the shot was fatal, he dragged his own body down to the lake. Pulling up a couple of cinder blocks, he secured these to his body with a piece of twine. After all, it only had to stay under the water for today. There would be no tomorrow; his bullet had killed the future.

It was heavy work. He had not realized how heavy he was, or how far out of shape he had gotten with age. By the time he had reached the water, he was already seriously out of breath; but he didn't have time to rest—no one must know that he had killed himself. Lifting the body again to secure the blocks to it, he wondered if he would be able to finish before the fireworks crew arrived. More than once he thought he heard a car or truck coming, but as yet it was just morning traffic. He had intended to lift himself onto his shoulders and throw himself several yards out—there was a deep spot right near the shore here where he used to fish, and the body would sink far enough that it couldn't be found. But he was not so strong as he believed, and already far more tired than he anticipated. He barely was able to shove the body into the water before he collapsed on the shore. He saw it slide through the mud out of sight before he closed his own eyes for the last time.

So it was that the fireworks company found the body of Dr. Arnold Hess lying on the shore of the lake when they arrived twenty minutes later. Although they arranged for the ambulance to remove him secretly, it was for them the beginning of the worst picnic of their lives.

About The Author

"M.J." is an avid writer and gamer, co-creator of *Multiverser: The Game*, and well known for his web site *Temporal Anomalies in Popular Time Travel Movies*; some of his writings have been translated into French and German. He can be found on Facebook and other social media platforms, and through *MJYoung.net*, *the Christian Gamers Guild*, and *Patreon*.

Looking for more fantastic books?
try
dimensionfold.com

www.ingramcontent.com/pod-product-compliance
Lightning Source LLC
Chambersburg PA
CBHW071431070526
44578CB00001B/70